PHARMACEUTICAL
BOTANY

HEBER. W. YOUNGKEN, Ph. G., A. M.

Assistant Professor of Botany and Pharmacognosy
at the Medico-Chirurgical College; Member of
the American Pharmaceutical Association,
American Association for the Advancement of Science, Etc.

MAVEN BOOKS

Chennai   Trichy   Tirunelveli   New Delhi

MAVEN BOOKS

An Imprint of **MJP Publishers**

ISBN 978-93-87867-75-8   **MAVEN Books**

All rights reserved   No. 44, Nallathambi Street,
Printed and bound in India   Triplicane, Chennai 600 005

MJP   752   © Publishers, 2019

Publisher :   **C. Janarthanan**

Publisher's Note

The legacy of a country is in its varied cultural heritage, historical literature, developments in the field of economy and science. The top nations in the world are competing in the field of science, economy and literature. This vast legacy has to be conserved and documented so that it can be bestowed to the future generation. The knowledge of this legacy is slowly getting perished in the present generation due to lack of documentation.

Keeping this in mind, the concern with retrospective acquiring of rare books has been accented recently by the burgeoning reprint industry. MAVEN Books is gratified to retrieve the rare collections with a view to bring back those books that were landmarks in their time.

In this effort, a series of rare books would be republished under the banner, "MAVEN Books". The books in the reprint series have been carefully selected for their contemporary usefulness as well as their historical importance within the intellectual. We reconstruct the book with slight enhancements made for better presentation, without affecting the contents of the original edition.

Most of the works selected for republishing covers a huge range of subjects, from history to anthropology. We believe this reprint edition will be a service to the numerous researchers and practitioners active in this fascinating field. We allow readers to experience the wonder of peering into a scholarly work of the highest order and seminal significance.

MAVEN Books

PREFACE

The aim has been to eliminate from this book all those topics that are of minor importance to the student and practitioner of Pharmacy. As a pharmacist and teacher, the writer feels that the botanical preparation for Pharmocognosy and Materia Medica, in those colleges where Botany is given for one year, should include mainly the structural and systematic aspects of the science. In the Medico-Chirurgical College, of Philadelphia, Botany is taught the first year, extending over a period of 155 hours. The author has introduced in this concise volume the important subject matter of his lectures given to first year students, and has omitted laboratory directions for the obvious reason that fixed subjects for laboratory study are unnecessary. It is not a book on Pharmacognosy, however, since it does not describe how one drug differs from another of the same group in all of its details.

The work is included in two parts. Part I is largely devoted to the morphology (gross and minute) and, to a less extent, the physiology of the Angiosperms. Part II deals with the taxonomy of plants, mainly but not wholly of medicinal value, together with the parts used and the names of the official and non-official drugs obtained from these.

The author does not claim sole originality for the facts presented, but has consulted many sources of information, mention of which will be found in the bibliography of the text.

Acknowledgment is here made to his esteemed friends, Dr. Francis E. Stewart of the Medico-Chirurgical College and Dr. John M. Macfarlane of the Univ. of Penna., for valuable assistance in the reading of the proofs and preparation of the index.

H. W. Y.

Philadelphia.

FOREWORD

In a monograph entitled "An Old System and a New Science," published in 1882, I advocated a return to the classification in which knowledge relating to the Materia Medica is embraced under the general head "Pharmacology"; in my address as Chairman of the Section on Materia Medica, Pharmacy and Therapeutics, delivered at the forty-seventh annual meeting (1896) of the American Medical Association, the same was again suggested; and in numerous papers on the subject since contributed to medical and pharmaceutical societies and press, the same plea was repeated. It is therefore gratifying to note the adoption of this classification by the National Committee Representing the Boards and Schools of Pharmacy of the United States for its "Pharmaceutical Syllabus," and also to note its incorporation into the New York State Pharmacy Law and adoption by the Board of Regents of the State of New York for the guidance of teachers of pharmacy in that state.

Pharmacology in its widest scope embraces the study of drugs from every possible point of view. As limited to the study of the changes incited in living organisms by the administration of drugs, we have excellent text books by Cushney, Sollman and others. But these works demand for their proper study more extended education than required by the national syllabus or the needs of the pharmaceutical student. The object of the Stewart Pharmacologic Manuals is to supply text books suitable for pharmacists and pharmaceutical colleges, and prepared in accordance with the national syllabus.

F. E. S.

CONTENTS

PART I

Terminology and Morphology

CHAPTER I

CHAPTER II

CONTENTS   xi

PART II

Taxonomy

ERRATA

Page 2 line 34 for "Cystology" read "Cytology."
Page 55 line 1 for "Ezalbuminous" read "Exalbuminous."
Page 60 Figs. 32 and 33 for *"Clasiceps"* read "Claviceps."
Page 83 line 21 for "pinnahfied" read "pinnatifid."
Page 91 line 22 for "prostate" read "prostrate."
Page 93 line 4 for "carolla" read "corolla."

TEXT-BOOK OF
PHARMACEUTICAL BOTANY

PART I

TERMINOLOGY AND MORPHOLOGY

DIVISIONS OF BOTANY

1. **Structural Botany** or **Plant Morphology** treats of the various organs or parts of a plant, as root, stem, flower, fruit, etc., with their special forms and modifications. It also includes **Vegetable Histology,** that part of structural botany which considers the minute or microscopical structure of plant tissues and **Vegetable Cytology,** which treats of plant cells and their contents.

2. **Physiological Botany** explains how the various parts of the plant perform their work of growth, reproduction and the preparation of food for the support of animal life from substances not adapted to that use.

3. **Geographical Botany** treats of the distribution of plant life on the globe. The centre of distribution for each plant is the habitat or original source from which it spreads, often over widely distant regions.

4. **Economic or Applied Botany** deals with the science from a practical standpoint, showing the special adaptation of the vegetable kingdom to the needs of everyday life.

5. **Geological Botany** treats of the plants of former ages, traceable in their fossil remains.

6. **Systematic Botany** or **Vegetable Taxonomy** considers the classification or arrangement of plants in groups or ranks according to their resemblances or differences.

7. **Vegetable Ecology** treats of plants in relation to their environment.

CLASSIFICATION OF PLANTS

By grouping together those plants which are in some respects similar and combining these groups with others, it is possible to form something like an orderly system of classification. Such a system based upon natural resemblances is called a **"natural system."**

Types represent general plans of structure.

A **Class** is formed by special modification of a type. Classes resembling each other are called **Series.**

An **Order** is a group of the same class, related by a common structure.

A **Family** is a group of the same order, related by a common structure.

A **Genus** is a still smaller group having the same essential structure.

A **Species** is the smallest group whose structure is constant.

An **Individual** is a unit of organic life, forming a complete animate existence.

A **Variety** is a peculiarity of **Race.** Races and varieties are both sub-divisions of species.

A **Hybrid** is a cross-breed of two varieties or species, rarely of two genera.

SUBDIVISIONS OF THE VEGETABLE KINGDOM

The two great sub-divisions of the vegetable kingdom are:

Phanerogams or flowering plants and **Cryptogams** or flowerless plants.

The Phanerogams are further divided into:

Angiosperms, characterized by having their seeds enclosed within a box-like covering.

Gymnosperms, which have their seeds borne naked. (They are polycotyledonous.)

The Angiosperms are classified according to the number of their cotyledons, or seed leaves in the embryo, into:

Monocotyledonous plants, which have one cotyledon, as Indian Corn and Ginger, and

Dicotyledonous plants, which have two cotyledons, as Burdock, and Ipecacuanha.

VEGETABLE CYSTOLOGY

CELLULAR STRUCTURE

The bodies of all plants are made up of one or more units of structure called cells.

A **cell** is a mass of protoplasm containing a nucleus.

Protoplasm is the more or less semi-fluid, viscid, foamy, and granular substance in which life resides. It is the "physical basis of life." Vegetable cells generally have cell walls of cellulose surrounding the living protoplasm of the cell (protoplast).

Cells divide to form tissues.

PROTOPLASMIC CELL CONTENTS

Protoplasm consists of four well-differentiated portions:

(a) **Cytoplasm,** or the foamy, often granular matrix of protoplasm outside of the nucleus.

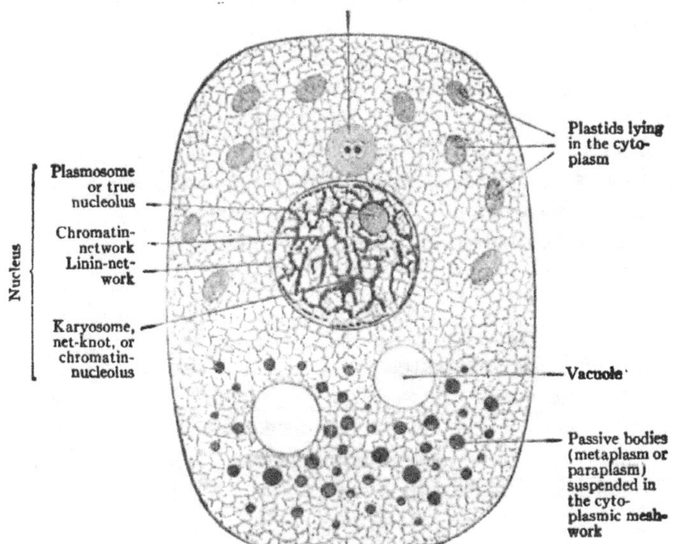

FIG. 1.—Diagram of a cell. (*From Hegner's Zoölogy, after Wilson, published by the Macmillan Co.*)

(b) **Nucleus or Nuclearplasm,** a denser region of protoplasm containing chromatin, a substance staining heavily with certain basic dyes.

(c) **Nucleolus,** a small body of dense protoplasm within the nucleus.

(d) **Plastids,** composed of plastid plasm, small discoid bodies scattered about in the cytoplasm. Sometimes, as in the cells of lower plants like the Spirogyra, plastids are large and are then called **chromatophores.**

According to the position of the cells in which plastids occur and the work they perform, they differ in color, viz.:

Leucoplasts are colorless plastids found in the underground portions of a plant and also in seeds, and the egg cell. Their function is to build up reserve starch from sugar and other carbohydrates as well as to change the reserve starch back into sugar when it is needed for the growth of the plant.

Chloroplasts are plastids found in cells exposed to light and contain the green pigment, chlorophyll.

Chromoplasts are plastids found in cells independent of their relation to light or darkness and contain the yellow or orange pigment called chromophyll.

During cell division another protoplasmic body appears called a **centrosome.**

NON-PROTOPLASMIC CELL CONTENTS

1. **Starch** { ASSIMILATION. / RESERVE. }

2. **Inulin.**

3. **Sugars** such as DEXTROSE, LEVULOSE, SACCHAROSE, MALTOSE, GENTIANOSE, MANNITOL.

4. **Cell-sap colors** (in solution of cell sap).

5. **Alkaloids.**

6. **Glucosides.**

7 **Neutral Principles.**

8. **Feebly basic substances.**

9. **Aleurone grains.**

10. **Calcium Oxalate** { ROSETTE AGGREGATES. / MONOCLINIC PRISMS. / CRYSTAL FIBRES. / RAPHIDES. / MICRO-CRYSTALS. }

11. **Cystoliths.**

12. **Tannin.**

13. **Gums and Mucilage.**

14. **Oils.**

15. **Resins.**

16. **Enzymes** { PROTEOLYTIC. / DIASTASES. / INVERTASES. }

PLANT TISSUES

A **tissue** is an aggregation of cells of common source, structure and function in intimate union.

According to structure the following tissues are found in various forms of higher plants:

<table>
<tr><td>1. MERISTEM</td><td>8. LATICIFEROUS TISSUE</td></tr>
<tr><td>2. PARENCHYME</td><td>9. CRIBIFORM OR SIEVE TISSUE</td></tr>
<tr><td>3. COLLENCHYME</td><td>10. WOODY FIBRE TISSUE</td></tr>
<tr><td>4. SCLERENCHYME</td><td>11. HARD BAST</td></tr>
<tr><td>5. EPIDERMIS</td><td>12. TRACHEARY TISSUE</td></tr>
<tr><td>6. ENDODERMIS</td><td>13. MEDULLARY RAYS</td></tr>
<tr><td>7. CORK</td><td></td></tr>
</table>

A mass of tissue so united in the plant as to constitute a distinct unit is called a **tissue system.** Three systems of tissues are commonly distinguished in higher forms of plants:

THE EPIDERMAL OR TEGUMENTARY SYSTEM
THE FUNDAMENTAL SYSTEM
THE FIBROVASCULAR SYSTEM

Meristem, frequently called embryonic tissue, is undifferentiated tissue composed of cells in the state of rapid division.

Parenchyme or **Fundamental Tissue** is the soft ground tissue of plants consisting of cells about equal in length, breadth and thickness (isodiametric) with thin cellulose cell walls enclosing protoplasm and a nucleus. Three important kinds of parenchyme, viz.: Chlorophyll parenchyme, containing chloroplasts and found in green parts of plants; reserve parenchyme occurring in seeds, and underground parts of plants and containing starch, aleurone grains, fixed oils, etc.; conducting parenchyme found distributed in various parts of plants and serving for the transferral of food.

Collenchyme consists of elongated prismatic cells whose walls are of cellulose. The angles of the cells are thickened with a colloidal substance. It is found beneath the epidermis of many plants, rarely alongside the endodermis and forms the "ribs" of stems such as in Burdock, Caraway, etc. Its function is that of support.

Sclerenchyme or "stony tissue" is made up of stone cells variously shaped. These were formerly parenchyme cells whose walls became thickened by deposits of lignin, layer within layer until the cells are often nearly filled with this substance. A lumen is found within the centre of a stone cell which is in communication with radial pore canals leading outward and in communication with the pore canals of adjacent stone cells. Stone cells are distributed in fruits as gritty particles, in barks and seeds. They are supporting structures.

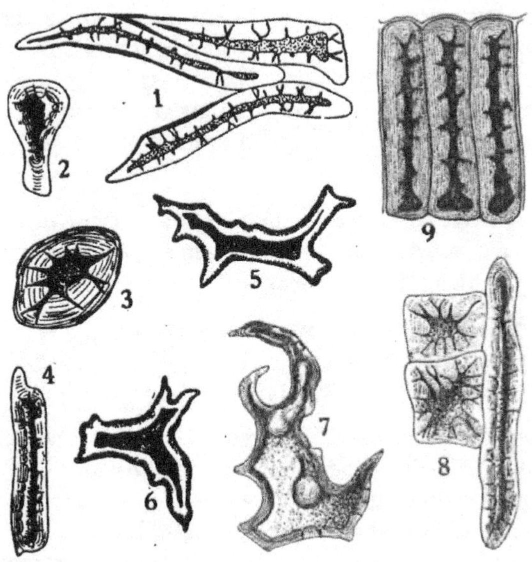

FIG. 2.—Stone cells from different sources. 1, From coffee; 2, 3 and 4, from stem of clove; 5 and 6, from tea leaf; 7, 8 and 9, from powdered star-anise seed. (*From Stevens after Moeller.*)

Epidermis is the outer covering tissue of a plant and is protective in function. Its cells may be brick-shaped, polygonal, equilateral or wavy in outline. Their outer walls are cutinized (infiltrated with a waxy-like substance called cutin). Among the epidermal cells of leaves and young green stems may be found numerous pores or stomata (sing. stoma) surrounded by pairs of crescent-shaped cells, called guard cells. The stomata are in direct communication with air chambers beneath them which in turn are in communication with intercellular spaces of the tissue beneath. The function of the stomata is to give off watery vapor and take in or give off carbon dioxide, water and oxygen. In addition to stomata some leaves possess groups of water stomata which differ from transpiration stomata in that they always remain open, are circular in outline, give off water in droplets directly, and lie over a quantity of small-celled glandular material which is in connection with one or more fibrovascular bundles.

Endodermis is the starch sheath layer of cells, constituting the innermost layer of cortex whose radial walls are more or less suberized.

Cork or suberous tissue is composed of cells of tabular shape, whose walls possess suberized layers. Its cells are mostly filled with air containing a yellow or brownish substance. It is derived from the phellogen or cork cambium which cuts off cork cells outwardly. Cork tissue is devoid of intercellular air spaces. It is protective in function.

Laticiferous tissue is to be seen in many groups of plants principal among which are the Asclepiadaceæ, Euphorbiaceæ, Apocynaceæ Urticaceæ and Papaveraceæ. Its cells are elongated, tubular, often branched and penetrate all the organs of plants in which they are found. They contain a milky-white or colored emulsion of gum-resins, fat, wax, caoutchouc, and in some cases, alkaloids, tannins, salts ferments, etc.

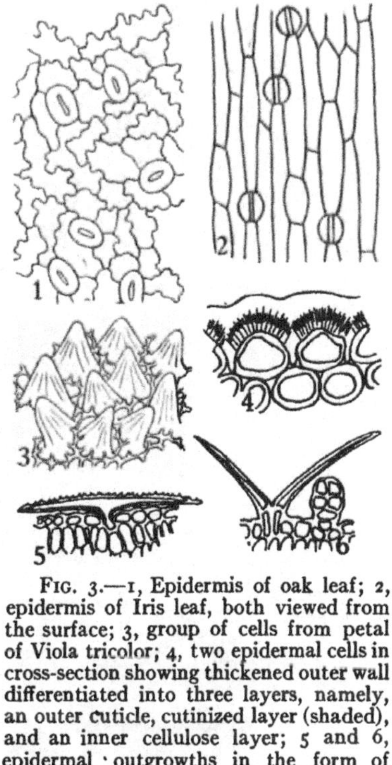

FIG. 3.—1, Epidermis of oak leaf; 2, epidermis of Iris leaf, both viewed from the surface; 3, group of cells from petal of Viola tricolor; 4, two epidermal cells in cross-section showing thickened outer wall differentiated into three layers, namely, an outer cuticle, cutinized layer (shaded), and an inner cellulose layer; 5 and 6, epidermal outgrowths in the form of scales and hairs. (3 *after Strasburger*, 4 *after Sachs, and 5 after de Bary*.)

Cribiform or Sieve tissue consists of superimposed, elongated, tubular, thin-walled cells whose transverse walls, called sieve plates, are perforated, permitting of the passage of albuminous substances from one cell to another.

Woody Fibres are elongated, thick-walled, and taper-ended lignefied elements found in the xylem region of the fibrovascular bundle accompanying the tracheæ (ducts). The walls of these fibres show oblique pores. Woody fibres are the supporting elements of the xylem.

Hard Bast is composed of elongated, spindle-shaped, thick-walled elements called bast fibres. The characteristic thickening of the walls

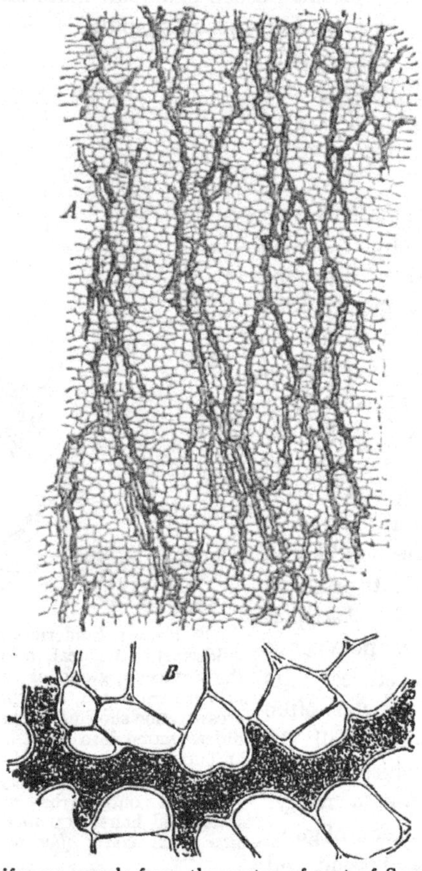

FIG. 4.—Laticiferous vessels from the cortex of root of Scorozonora hispanica. *A*, as seen under low power, and *B*, a smaller portion under high power. (*From Stevens after Sachs.*)

of these fibres is due to deposits of lignin upon the inner surface of the cellulose cell wall. Like the woody fibres the lumina of these contain air and the fibre walls are provided with oblique pores. Bast is the supporting tissue of the phloem.

Tracheary tissue consists of tracheæ (ducts or vessels) and tracheids, both of which are found in the xylem region of the fibrovascular bundle and have as their function the conduction of water with mineral salts in solution from the roots upward. The tracheæ or ducts are elongated, slightly lignefied tubes with occasional cross-walls and having characteristic thickenings on their inner surface. Tracheæ are classified as:

ANNULAR, with ring-like thickenings.

SPIRAL, with spiral thickenings.

RETICULATE, with reticulate thickenings.

POROUS or PITTED with spherical or oblique slit pores.

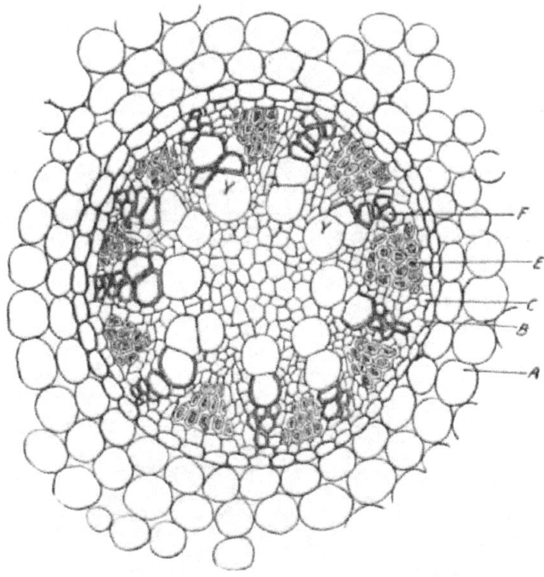

FIG. 5.—Cross-section through a portion of a root of *Acorus calamus*. *A*. Cortical parenchyma; *B*. endodermis; *C*. pericycle; *E*. phloëm *F*. xylem. At *Y, Y*, are large tracheal tubes, which were formed last, the narrow tubes near the periphery of the xylem being formed first. At the center of the root, within the circle of vascular bundles, occur thin-walled parenchymatous pith cells. (*From Sayre after Frank.*)

Tracheids are undeveloped ducts having bordered pores and frequently scalariform thickenings.

Medullary Rays are bands of parenchymatous cells extending radially from the cortex to the pith (primary med. rays) or from a part of the xylem to a part of the phloem (secondary med. rays).

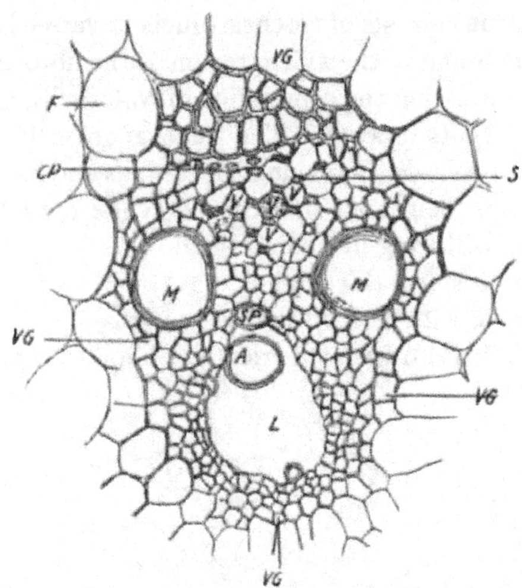

FIG. 6.—Closed bundle of stem of *Zea mays*. *VG*, Bundle sheath; *L*, intercellular space; *A*, ring from an annular tracheal tube; *SP*, spiral tracheal tube; *M*, pitted vessels; *V*, sieve tubes; *S*, companion cells; *CP*, crushed primary sieve tubes; *F*, thin-walled parenchyma of the ground or fundamental tissue. (*From Sayre after Strasburger.*)

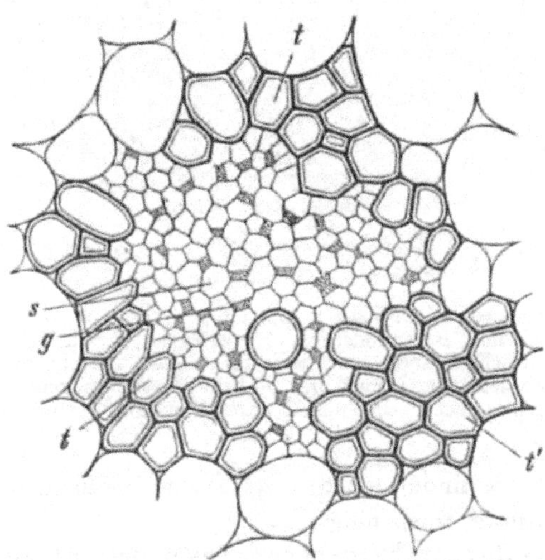

FIG. 7.—Transverse section of a concentric bundle from the rhizome of *Iris*. Xylem surrounding the phloëm. *t*, Tracheæ; *t¹*, protoxylem; *s*, sieve tubes; *g*. companion cells of the internal phloëm portion. (*From Sayre after Vines.*)

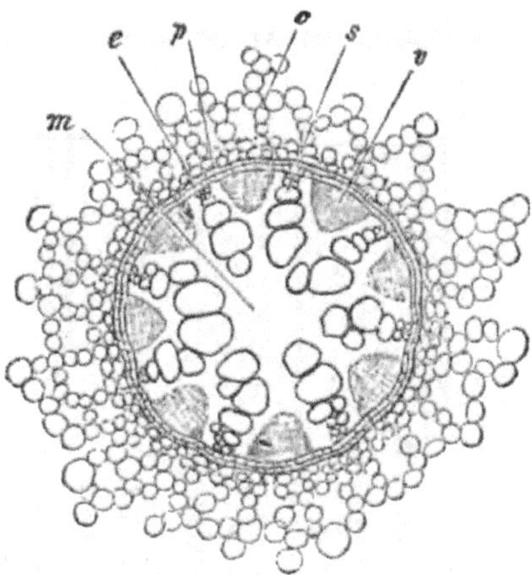

FIG. 8.—Transverse section of central part of the root of *Acorus calamus*. *c*, Cortex; *e*, endodermis; *p*, pericycle; *s*, primary xylem or wood-bundles, with small spiral vessels of the protoxylem externally; *v*, phloëm portion of vascular bundles; *m*, pith. (*From Sayre after Strasburger.*)

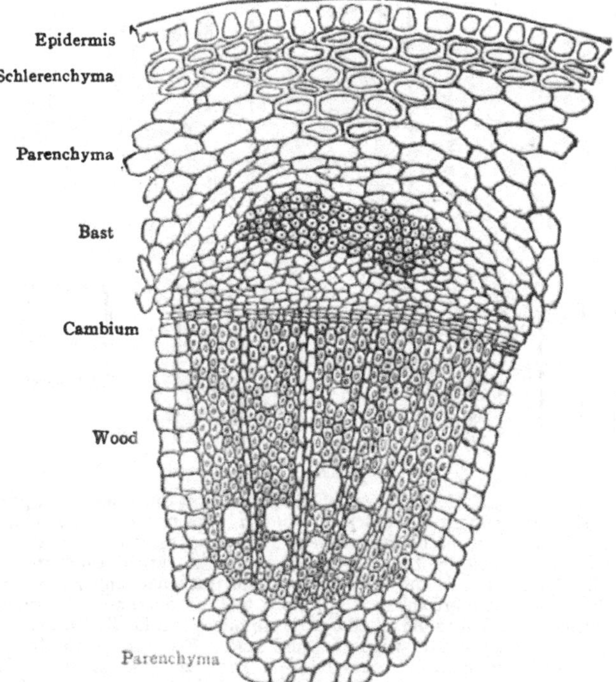

FIG. 9.—Cross-section of a typical dicotyledon stem from the pith to the epidermis and comprising one vascular bundle. (*From Hamaker.*)

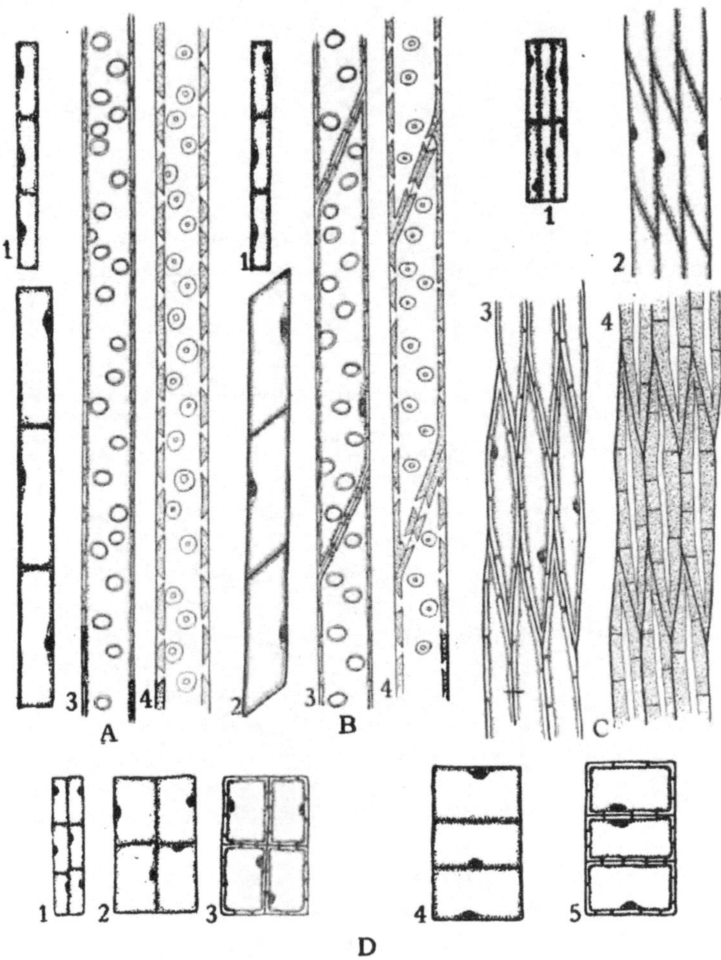

FIG. 10.—Stages in the development of the elements of the xylem. *A*, progressive steps in the development of a tracheal tube. 1, Row of procambial or cambial cells that are to take part in the formation of a tube; 2, the same at a later stage enlarged in all dimensions; 3, the cells in 2 have grown larger, their cross-walls have been dissolved out, and the wall has become thickened and pitted; 4, the walls in 3 have become more thickened, the pits have an overhanging border, the walls have become lignified as indicated by the stippling, and finally the protoplasts have disappeared, and the tube is mature and dead. *B*, Stages in the formation of tracheids from procambial or cambial cells. The steps are the same as in *A*, excepting that the cross-walls remain and become pitted. *C*, steps in the development of wood fibers from cambial cells. 1, Cambial cells; 2, the same growth larger in all dimensions with cells shoving past each other as they elongate; 3, a

Fibrovascular Bundles are groups of fibres, vessels and cells coursing through the various organs of a plant and serving for conduction and support. According to the relative structural arrangement of their xylem and phloem masses they may be classed as follows:

I. CLOSED COLLATERAL, consisting of a mass of xylem lying alongside of a mass of phloem, the xylem facing toward the centre, the phloem facing toward the exterior. Stems of most Monocotyledons and Horsetails.

II. OPEN COLLATERAL, consisting of a mass of xylem facing toward the pith and a mass of phloem facing toward the exterior and separated from each other by a cambium. Stems and leaves of Dicotyledons and roots of Dicotyls and Gymnosperms of secondary growth.

III. BICOLLATERAL, characterized by a xylem mass being between an inner and an outer phloem mass. There are two layers of cambium cells, one between the xylem and inner phloem mass, the other between the xylem and outer phloem mass. Seen chiefly in stems of the *Cucurbitaceæ*.

IV. CONCENTRIC, characterized by a central xylem mass surrounded by a phloem mass or vice versa. No cambium present.

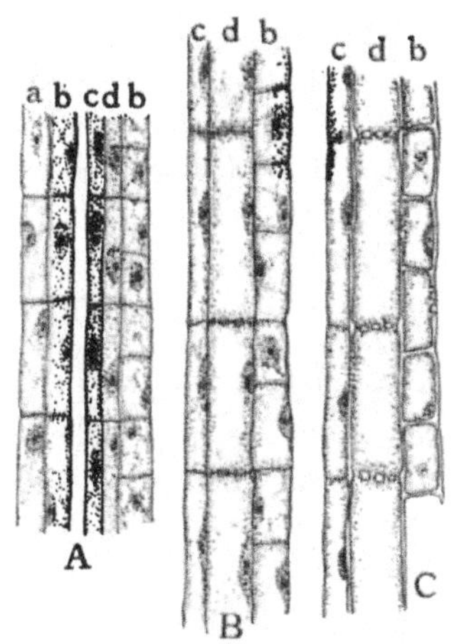

FIG. 11.—Stages in the development of sieve tubes, companion cells, and phloëm parenchyma. *A*, *a* and *b*, Two rows of procambial cells; in *c* and *d*, *a* has divided longitudinally and *c* is to become companion cells; *d*, a sieve tube, and *b*, phloëm parenchyma. *B*, *c*, Companion cells, and *d*, a beginning sieve tube from *c* and *d*, respectively in *A*. The cross-walls in *d* are pitted; *b*, phloëm parenchyma grown larger than in *A*. *C*, The same as *B* with the pits in the cross-walls of the sieve tubes become perforations, and the nuclei gone from the cells composing the tube. (*From Stevens.*)

later stage with cells longer and more pointed and walls becoming thickened and pitted; 4, complete wood fibers with walls more thickened than in the previous stage and lignified, as shown by the stippling. The protoplasts in this last stage have disappeared and the fibers are dead. *D*, steps in the formation of wood parenchyma from cambial or procambial cells. 1, Group of cambrial or procambial cells; 2, the same enlarged in all dimensions; 3, the same with walls thickened and pitted; 4 and 5 show the same stages as 2 and 3, but here the cells have enlarged radially or tangentially more than they have vertically. The walls of these cells are apt to become lignified, but the cells are longer lived than the wood fibers. (*From Stevens.*)

(*a*) *Concentric*, with xylem central in bundle. Seen in stems and leaves of nearly all ferns and the *Lycopodiaceæ*.

(*b*) *Concentric*, with phloem central in bundle. Seen in stems and leaves of some *Monocotyledons*. Ex.: Calamus.

V. RADIAL, characterized by a number of xylem and phloem masses alternating with one another. Seen in the roots of all *Spermatophytes* and *Pteridophytes*.

XYLEM is that part of a fibrovascular bundle that contains wood cells and fibres. It may also contain tracheæ, tracheids, seldom sieve tubes.

PHLOEM is that part of a fibrovascular bundle that contains sieve tubes, phloem cells, and often bast fibres.

Classification of Tissues According to Function.—According to their particular function, tissues may be classified as follows:

I. CONDUCTING TISSUE
- Parenchyme (Fundamental tissue).
- Xylem cells.
- Tracheæ (ducts).
- Phloem cells.
- Sieve tubes.

II. PROTECTIVE TISSUES
- Epidermis (outer cell walls cutinized)
- Cork (suberized tissue).

III. MECHANICAL TISSUES
- Bast fibres.
- Wood fibres.
- Stone cells.

PLANT ORGANS AND ORGANISMS

An *organ* is a part of an organism made up of several tissues and capable of performing some special work.

An *organism* is a living entity composed of different organs or parts with functions which are separate, but mutually dependent, and essential to the life of the individual.

The organs of flowering plants are either **Vegetative** or **Reproductive.** The vegetative organs of higher plants are PLANT HAIRS, ROOTS, STEMS, and LEAVES. They are concerned in the absorption and elaboration of food materials either for tissue-building or storage.

The reproductive organs of higher plants include those structures whose function it is to continue the species, viz., the flower, fruit and seed.

The ripened seed is the product of reproductive processes, and the starting point in the life of all Phanerogams. The living part of the seed is the *embryo*, which, when developed, consists of four parts, the caulicle, or rudimentary stem, the lower end of which is the beginning of the root, or radicle. At the upper extremity of the stem are two thickened bodies, closely resembling leaves, known as cotyledons, and between these a small bud or plumule.

The function of the cotyledon is to build up nourishment for the rudimentary plantlet until it develops true leaves of its own.

THE ROOT

The root is that part of the plant that grows into or toward the soil, that never develops leaves, rather rarely' produces buds, and whose growing apex is covered by a cap.

The functions of a root are absorption, storage and support. Its

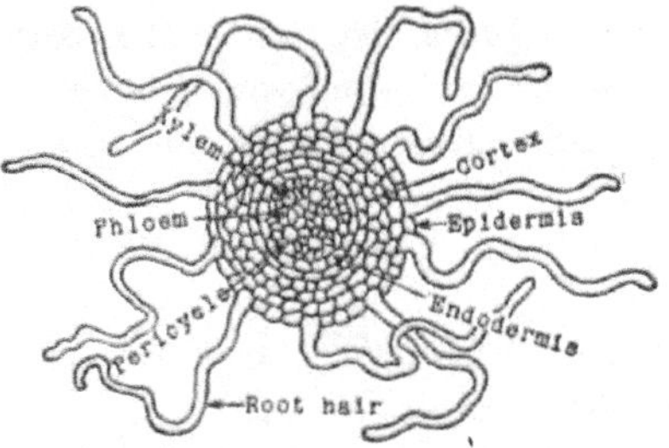

Fig. 12.—Cross-section of rootlet in the region of the root hairs. (*From Stevens.*)

principal function is the absorption of nutriment and to this end it generally has branches or rootlets covered with root hairs which largely increase the absorbing surface. These root hairs are of minute and simple structure, being merely elongations of the epidermis of the root back of the root cap into slender tubes with thin walls.

The tip of each rootlet is protected by a sheath- or scale-like covering known as the ROOT CAP, which not only protects the delicate growing point, but serves as a mechanical aid in pushing its way through the soil. The generative tissues in the region of the root cap are: PLEROME, producing fibrovascular tissue; PERIBLEM, producing cortex; DERMATOGEN, producing epidermis; and CALYPTROGEN, producing the root cap.

DIFFERENCES BETWEEN ROOT AND STEM

The Root	*The Stem*
1. Descending axis of plant.	1. Ascending axis of plant.
2. Growing point sub-apical.	2. Growing point apical.
3. Contains no chlorophyll.	3. Chlorophyll sometimes present.
4. Branches arranged irregularly.	4. Branches with mathematical regularity.
5. Does not bear leaves or leaf rudiments.	5. Bears leaves and modifications.
6. Structure comparatively simple.	6. Structure better defined.

Classification of Roots as to Form.—1. PRIMARY or FIRST ROOT, a direct downward growth from the seed, which, if greatly in excess of the lateral roots, is called the MAIN or TAP ROOT. Ex.: Taraxacum, Radish.

2. SECONDARY ROOTS are produced by the later growths of the stem, such as are covered with soil and supplied with moisture. Both primary and secondary roots may be either fibrous or fleshy.

The grasses are good examples of plants having fibrous roots. Fleshy roots may be multiple, as those of the Dahlia, or may assume simple forms, as follows:

Fusiform, or *spindle-shaped*, like that of the radish or parsnip.

Napiform or *turnip-shaped*, somewhat globular and becoming abruptly slender then terminating in a conical tap root, as the roots of the turnip.

Conical, having the largest diameter at the base then tapering, as in the Maple.

3. ANOMALOUS ROOTS are of irregular or unusual habits, subserving other purposes than the normal.

4. ADVENTITIOUS ROOTS are such as occur in abnormal places on the plant. Ex.: Roots developing on Bryophyllum leaves.

5. EPIPHYTIC ROOTS, the roots of epiphytes, common to tropical forests, for example, never reach the soil at all, but cling to the bark of trees and absorb nutriment from the air. Ex.: Roots of Vanilla.

6. The roots of parasitic plants are known as HAUSTORIA. These penetrate the bark of plants upon which they find lodgment, known as hosts, and absorb nutritious juices from them. The *Mistletoe, Dodder and Geradia* are typical parasites.

Duration of Root.—Plants are classified according to the duration of the root, as follows:

1. ANNUAL PLANTS are *herbs* with roots containing no nourishment for future use. They complete their growth, producing flower, fruit and seed in a single season, then die.

· 2. BIENNIAL plants develop but one set of organs the first year, and as in the beet and turnip, etc., a large amount of reserve food material is stored in the root for the support of the plant the following season when it flowers, fruits, and dies.

3. PERENNIAL PLANTS live indefinitely, as trees.

Root Histology. MONOCOTYLEDONS.—The histology of monocotyledonous roots varies, depending upon relations to their surroundings, which may be aquatic semi-aquatic, mesophytic, or xerophytic.

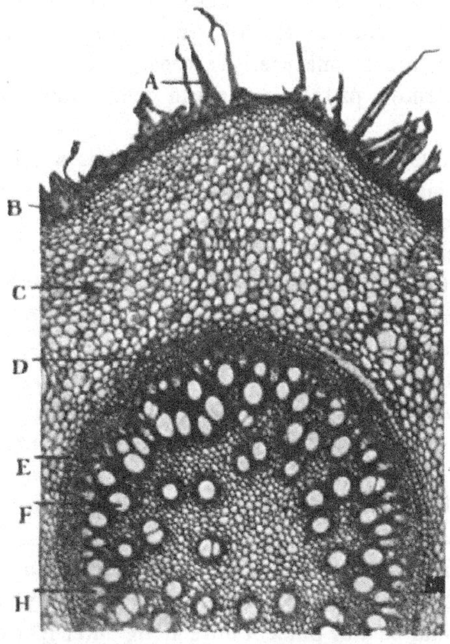

FIG. 13.—Sarsaparilla, Mexican. Cross-section of root. (32 diam.) *A*, Root hairs; *B*, cork; *C*, parenchyma of cortex; *D*, endodermis; *E*, wood parenchyma and fibers; *F*, water tube; *H*, phloëm. (Photomicrograph.) (*From Sayre.*)

In this connection we will discuss only the type of greatest pharmacognic importance, *i.e.*, the mesophytic type as seen in its most typical form in the transverse section of an Onion root.

Examining such a section from periphery toward the centre, one notes the following:

1. Epidermis with thin cuticle.

2. Cortex, consisting of broad zone of rounded cells getting larger

and then smaller in calibre.   These store starch and enable sap to pass through.

3. Endodermis, or innermost layer of cells in the cortex with lenticularly thickened radial walls.

4. Pericambium, a zone of one, two, or three layers of rounded, thin-walled, actively dividing cells, which may give rise to side rootlets.

5. Radial fibrovascular bundle, which in most monocotyledons consists of eight, twelve, or fifteen alternating patches of phloem with radiating xylem arms between.   Phloem tissue consists of phloem cells and sieve tubes.   Xylem at tips of arms, made of spiral tracheæ the first xylem elements to mature.   Internal to these are small pitted vessels, later, striking pitted vessels and considerable wood fibre.

6. Pith.

DICOTYLEDONS.—The typical dicotyl root is a tetrarch one, four xylem alternating with four phloém patches.   These roots have an unlimited power of growth.

A. *Of Primary Growth.*

A trans-section of a dicotyl root in its young growth shows the following structure from periphery toward centre.

1. Epidermis with cutinized outer walls.

2. Hypodermis.

3. Cortex with usually small intercellular spaces.

4. Endodermis, or innermost layer of cells of the cortex with radially thickened walls.

5. Pericambium of one to two layers of actively growing cells which may produce side rootlets.

6. Radial fibrovascular bundle of four, rarely two or three or five or six phloem patches alternating with as many xylem arms.   Not uncommon to find bast or phloem fibre along outer face of each phloem patch.   Xylem has spiral tracheæ, internal to these a few pitted vessels.   Then, as root ages, more pitted vessels, also xylem cells and wood fibres make their appearance

B. *Of Secondary Growth.*   (Most official roots.)

At about six weeks one notes cells dividing by tangential walls in the inner curve of phloem patches.   This is intrafascicular cambium. A single layer of flattened cells starts to cut off on its inner side a quantity of secondary xylem and pushes out the patches of bast fibres, adds a little secondary phloem on the outer side.   Secondary xylem finally fills up the patches between the arms.   The patches of bast fibres get flattened out. The pericambium has a tendency to start division into an inner and outer layer. The outer layer becomes a cork cambium (phellogen) surrounding the bundle inside the endodermis It cuts off cork tissue on its outer face, hence all liquid material is prevented from filtering in and cortex including endodermis, as well as the epidermis, shrivel and dry up and separate off at the age of two to three months. The cork cambium (phellogen) may lay down secondary cortex internal to itself and external to the pholem.

Patches of cells of the *inner layer* of *pericambium* divide rapidly and are called interfascicular cambium. These join the intrafascicular cambium to form a continuous cambium ring which then cuts off additional secondary xylem on its inner face. and secondary phloem on its outer face pushing inward the first-formed or protoxylem and outward the first-formed or protophloem.

THE BUD

Buds are rudimentary stems with rudimentary leaves compactly arranged upon them.

The COTYLEDONS and PLUMULE represent the first bud on the initial stem or caulicle.

SCALY BUDS are such as have their outer leaf rudiments transformed into scales, often coated with a waxy or resinous substance without and a downy lining within, to protect them from sudden changes in climate. Ex.: Hickory.

NAKED BUDS are those whose leaf rudiments are destitute of coverings.

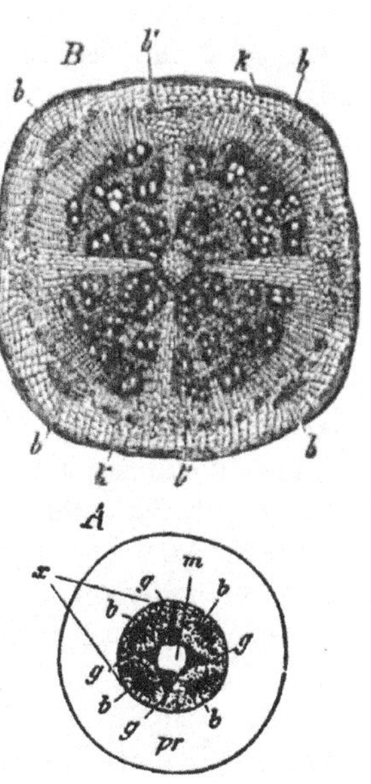

Fig. 14.—Cross-section of a young root of Phaseolus multiflorus. *A*, *pr*, cortex; *m*, pith; *x*, stele (all tissues within the endodermis collectively); *g*, *g*, *g*, *g*, primary xylem bundle; *b*, *b*, *b*, *b*, primary phloem bundle. *B*, cross-section through older portion of root of the same plant. *b′*, *b′*, Secondary bast; *k*, *k*, periderm. The remaining letters stand for the same tissues as in *A*. Notice that the cambium has laid down medullary rays in front of the primary xylem, but has made secondary xylem behind the primary phloëm. (*From Stevens after Vines.*)

Leaf buds develop leaves.

Flower buds are unexpanded blossoms.

Mixed buds contain both flower and foliage.

As to position buds are either *terminal* or *axillary*, either located at the apex of the stem or branch or in the axils of the leaves. If they occur on other situations on the stem, or upon roots or leaves they are termed *adventitious buds*. If, as often happens, more than one bud forms in or near the axil of the leaf, it is called an *accessory bud*.

The Stem

The stem is that part of the plant axis which bears leaves or modifications of leaves and its branches are usually arranged with mathematical regularity.

The functions of a stem are to bear leaves or branches, connect roots with leaves, and conduct sap.

When the stem rises above ground and is apparent, the plant is said to be *caulescent*.

When no stem is visible, but only flower or leaf stalks, the plant is said to be *acaulescent*.

Stems vary in size from scarcely one-twenty-fifth of an inch in length, as in certain mosses, to a remarkable height of 400 ft. upward. The giant Sequoia of California attains the height of 420 ft.

Direction of Stem Growth.—Generally the growth of the stem is erect. Very frequently it may be:

Ascending, or rising obliquely upward.

Reclining, or at first erect but afterward bending over and trailing upon the ground. Ex.: Raspberry.

Procumbent, lying wholly upon the ground.

Decumbent, when the stem trails and the apex curves upward. Ex.: Vines of the Cucurbitaceæ.

Repent, creeping upon the ground and rooting at the nodes, as the Strawberry.

Stem Elongation.—At the tip of the stem there is found a group of very actively dividing cells (meristem) which is the growing point of the stem. All the tissues of the stem are derived from the cells of the growing point whose activity gives rise in time to three generative regions which are from without, inward:

(1) Dermatogen, forming epidermis;

(2) PERIBLEM, forming the cortex; and

(3) PLEROME, forming the fibrovascular elements.

Duration of Stems.—HERBACEOUS, dying down to the ground at the close of the season.

ANNUAL, an herb whose life terminates with the season.

BIENNIAL, where the stem dies at the end of the first season, the underground parts perfecting themselves and retaining their vitality to the next season, when seeds are produced and the plant dies completely.

PERENNIAL, when the underground parts retain their vitality indefinitely.

Above-ground Stems.—A TWINING stem winds around a support, as the stem of a bean or Morning Glory.

A CULM is a jointed stem of the Grasses and Sedges.

A CLIMBING or scandent stem grows upward by attaching itself to some support by means of aerial rootlets, tendrils or petioles.

The SCAPE is a stem rising from the ground and bearing flowers but no leaves, as the dandelion, violet, or blood root.

A TENDRIL is a modification of some special organ, as of a leaf stipule or branch, capable of coiling spirally and used by a plant in climbing. Present in the Grape, Pea, etc.

A SPINE or thorn is the indurated termination of a stem tapering to a point, as the thorns of the Honey Locust.

PRICKLES are outgrowths of the bark only and are seen in the roses.

A STOLON is a prostrate branch, the end of which, on coming in contact with the soil, takes root, so giving rise to a new plant. Ex.: Currant and Raspberry.

An UNDERSHRUB or SUFFRUITCOSE stem is a stem of small size and woody only at the base.

A SHRUBBY or FRUITCOSE stem is a woody stem larger than the preceding and freely branching near the ground.

HERB AND TREE

A TREE is a perennial woody plant of considerable size (20 ft. or more in height) and having as the above-ground parts a trunk and a crown of leafy branches.

An HERB is a plant whose stem does not become woody and permanent, but dies, at least down to the ground, after flowering.

3

Underground Stems.—A RHIZOME is a creeping underground stem, more or less scaly, sending off roots from its lower surface and stems from [its upper. The rhizome grows horizontally, vertically or obliquely, bearing a terminal bud at its tip. Its upper surface is marked with the scars of the bases of aerial stems of previous years.

The TUBER is a short and excessively thickened underground stem, borne usually at the end of a slender, creeping branch, and having numerous eyes or buds. Ex.: Tubers of the Potato, Aconite, and Jalap.

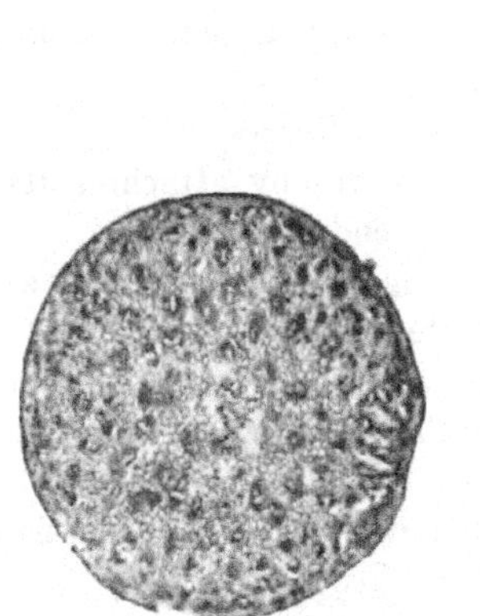
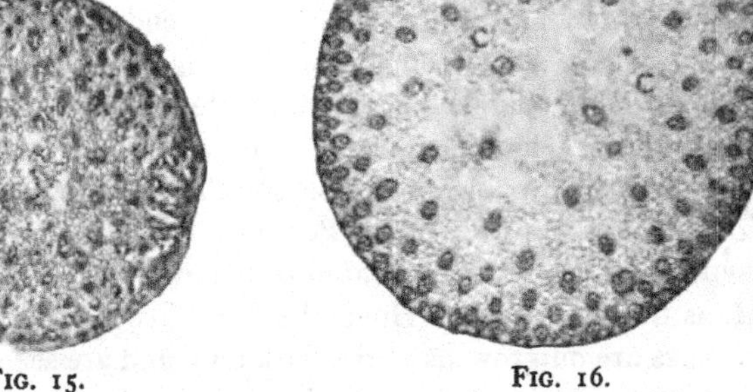

FIG. 15.   FIG. 16.

FIG. 15.—Photomicrograph of cross-section of very young cornstalk, where the procambium strands have just gone over into vascular bundles. For comparison with Fig. 16. (*From Stevens.*)

FIG. 16.—Photomicrograph of cross-section of cornstalk somewhat older than in Fig. 15. Compare with Fig. 15, and notice that the number of vascular bundles is approximately the same in both, and the number of cells in the fundamental tissue is approximately the same. Growth in Fig. 16 has been accomplished by the enlargement of the cells already present in Fig. 15. *a*, Epidermis; *b*, cortex and pericycle; *c*, *c*, fundamental or ground tissue corresponding to pith and medullary rays with vascular bundles interspersed through it. (*From Stevens.*)

The CORM is an underground stem excessively thickened and solid and characterized by the production of buds from the centre of the upper surface and rootlets from the lower surface.

A BULB is a very short and scaly stem, producing roots from the lower face and leaves from the upper.

TUNICATED BULBS are completely covered by broad scales which form concentric coatings. Ex.: Onion, Squill, Daffodil.

Scaly bulbs have narrow imbricated scales, the outer ones not enclosing the inner. Ex.: Lily.

Tubers and corms are annual. Bulbs and Rhizomes are perennial.

Exogenous and Endogenous Stems.—Exogenous stems are typical of Gymnosperms and Dicotyledons and can increase materially in thickness due to presence of a cambium. Such stems show differentiation into an outer or cortical region and an inner or central cylinder region.

Endogenous stems are typical of Monocotyledons and cannot increase materially in thickness due to absence of cambium. Such stems show no differentiation into cortical and central regions.

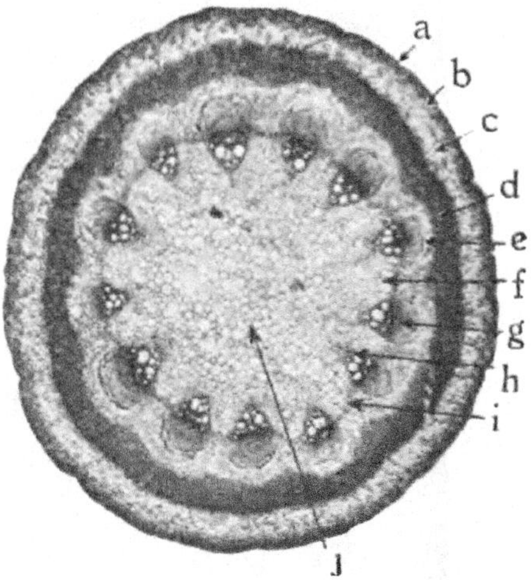

Histology of Annual Dicotyl Stem.—(In both annual and perennial dicotyledonous stems endodermis and pericambium are rarely seen since each has become so similar to cortex through passage of food, etc.)

Fig. 17.—Photomicrograph of cross-section of stem of Aristolochia sipho, where cambial activity is just beginning. a, Epidermis; b, collenchyma; c, thin-walled parenchyma of the cortex, the innermost cell layer of which is the starch sheath or endodermis; d, sclerenchyma ring of the pericycle; e, thin-walled parenchyma of the pericycle; f, primary medullary ray; g, phloëm; h, xylem; i, interfascicular cambium; j, medulla or pith. ×20. (*From Stevens.*)

1. Epidermis, cutinized, with hairs.

2. Cortex composed of three zones: an outer or exocortex, whose cells are thin walled and contain chloroplasts; a middle or mediocortex, consisting of cells of indurated walls giving extreme pliability and strength, an inner or endocortex, a very broad zone of thin and thicked-walled parenchyme cells.

3. The innermost layer of cells of the cortex called the endodermis. (Not generally distinguishable.)

4. Pericambium. (Not generally distinguishable.)

5. Fibrovascular bundles of open collateral type arranged in a circle with primary medullary rays between the bundles.

6. Pith.

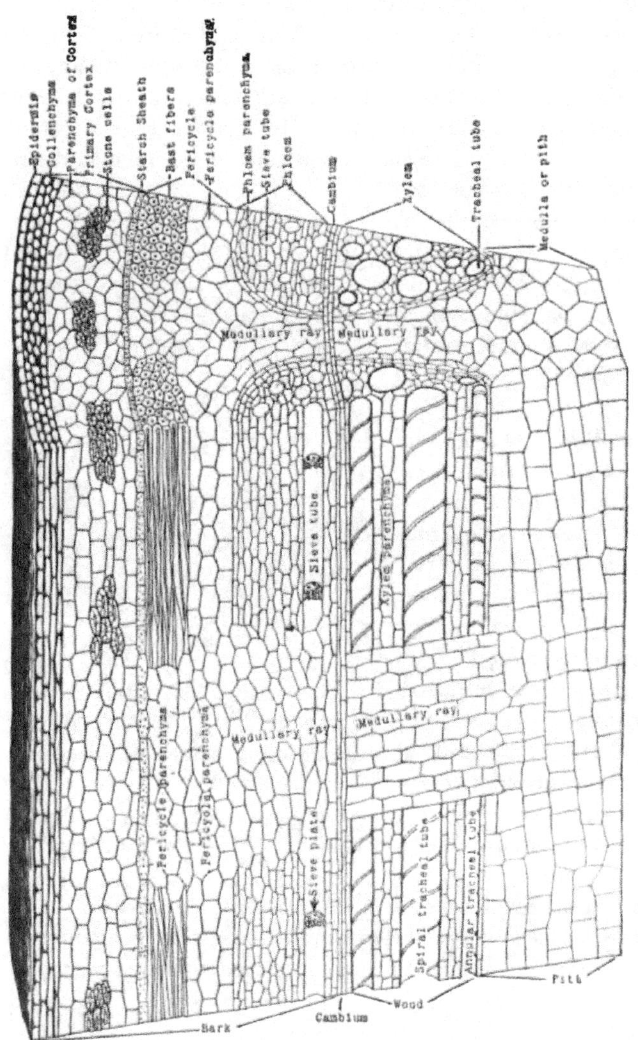

Fig. 18.—A diagram to show the character of the tissues and their disposition in a young stem of the typical dicotyledon type. (*From Stevens.*)

Growth of Perennial Dicotyl Stem and its Histology.—A perennial dicotyl stem in the first year does not differ in structure from an annual. By the close of the year a cork cambium (phellogen) has originated be-

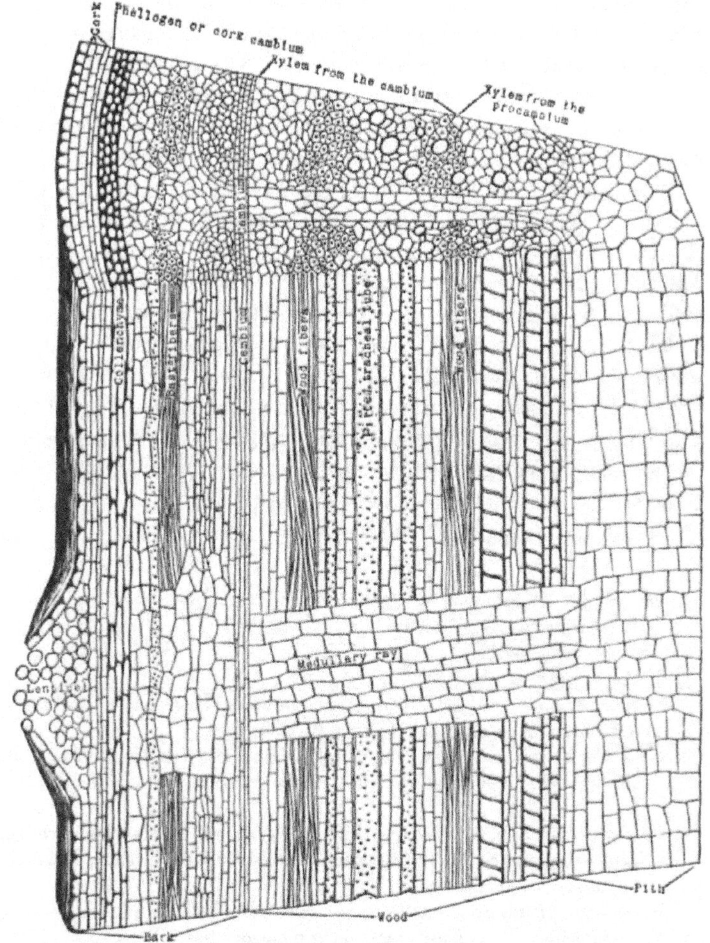

Fig. 19.—Diagram similar to the preceding but representing a later stage and showing the tissues formed by the cambium. (*From Stevens.*)

side the epidermis.   In origin of cork cambium—one of two methods: (a) either the epidermis may divide into an outer layer of cells that remains epidermis and an inner layer of cells that becomes cork cambium, or, (b) the outermost layer of cortex cells underneath the epidermis becomes active after being passive for one year, and lays down walls, the inner layer becoming cork cambium, the outer becoming a layer of cork. The cork cuts off water and food supplies from epidermis outside and so epidermis separates and falls off as stringy layer. The cork cambium produces cork on its outer face and secondary cortex on its inner.

Between the bundles certain cells of the primary medullary rays become very active and form interfascicular cambium which joins the cambium of the first-formed bundles (intrafascicular cambium) to form a complete cambium ring. By the rapid multiplication of these cambial cells new (secondary) xylem is cut off internally and new (secondary) phloem externally, pushing inward the first-formed, or PROTOXYLEM, and outward the first-formed, or PROTOPHLOEM, thus increasing the diameter of the stem. The primary medullary rays are deepened. Cambium may also give rise to secondary medullary rays.

Sometimes, as in Grape Vines, Honeysuckles, and Asclepias, instead of cork cambium arising from outer cortex cells it may arise at any point in cortex. It is the origin of cork cambium at varying depths that causes extensive sheets of tissue to separate off. That is what gives the stringy appearance to the stems of climbers.

At close of first year in *Perennial Dicotyl Stem* we note:

Periblem development

1. Epidermis—development of dermatogen or periblem—in process of peeling off, later on entirely absent.
2. Cork tissue or periderm.
3. Cork cambium or phellogen.
4. Sometimes zone of thin-walled cells containing chloroplasts cut off by cork cambium on inner face known as phelloderm.
5. Cortex—in perennial stem cells of cortex *may* undergo modification into mucilage cells, into tannin receptacles, crystal cells, spiral cells, etc.

Plerome development

6. Fibrovascular bundles of open collateral type which are now arranged into a compact circle, and between which are found primary and often secondary medullary rays.

From without inward the following tissues make up f. v. bundles.

Protophloem   { Hard Bast—long tenacious bast fibres.
Secondary Phloem   { Soft Bast—phloem cells and sieve tubes.

Cambium—active layer giving rise to secondary phloem on outer and secondary xylem on inner face, and adding to depth of med. rays.

Secondary xylem—wood fibres, pitted vessels, tracheids.

Protoxylem—spiral tracheæ.

Pith.

Lenticels and Their Formation —The epidermis in a great majority of cases produces stomata, apertures, surrounded by a pair of guard

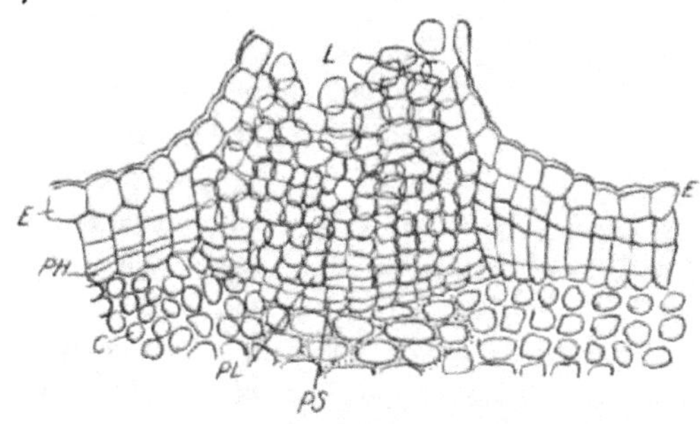

FIG. 20.—Cross-section through a lenticel of *Sambucus nigra*. *E*, Epidermis; *PH*, phellogen; *L*, loosely disposed cells of the lenticel; *PL*, cambium of the lenticel; *PS*, phelloderm; *C*, cortical parenchyma containing chlorophyll. (*From Sayre after Strasburger*.)

cells, which function as passages for gases and watery vapor from and to the active cells of the cortex beneath.

There very early originate in the region beneath the stomata loosely arranged cells from cork cambium which swell up during rain and rupture, forming convex fissures in the cork layer, called lenticels.

The function of lenticels is similar to that of stomata, namely, to permit of aëration of delicate cells of the cortex beneath.

Annual Thickening.—In all woody exogenous stems such as trees and shrubs the persistent cambium gives rise to secondary xylem thickening every spring, summer and autumn. Soon a great cylinder of xylem arises which constitutes the wood of the trunk and branches.

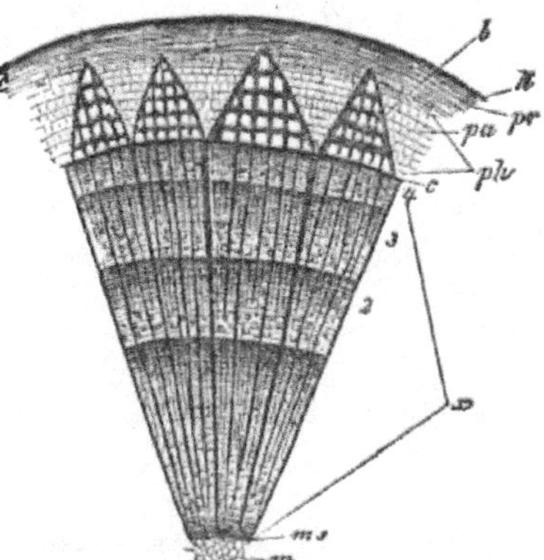

FIG. 21.—Part of a transverse section of a twig of the lime, four years old. *m*, Pith; *ms*, medullary sheath; *x*, secondary wood; *Ph*; phloëm. 2, 3, 4, annual rings; *c*, cambium; *pa*, dilated outer ends of medullary rays; *b*, blast; *pr*, primary cortex; *k*, cork. (*From Sayre after Vines*.)

In the spring, growth is more active, and large ducts with little woody fibre are produced while in summer or autumn growth is

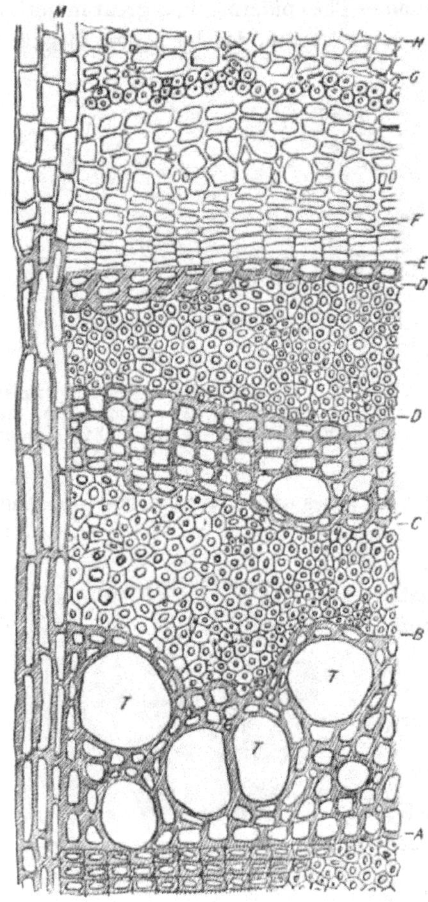

Fig. 22.—Part of a cross-section through branch of *Cytisus loburnum*. (The branch was cut from the tree at the end of October.) From *A* to *E* the last annual ring of wood; from *A* to *B* the spring growth with large tracheal tubes (*T, T, T*); between *B* and *C* and *D* and *D* are wood-fibers; between *C* and *D* and *D* and *E*, wood parenchyma; from *E* to *F*, canbium; *F* to *G*, p l ëm portion; *G* to *H*, cortical parenchyma; *M* medullary ray. Below *A* the last wood-fibers and wood parenyma formed the previous year. (*From Sayre after Haberlandt.*)

lessened and small ducts and much mechanical woody fibre are formed. Thus the open, loosely arranged product of the spring growth abuts on the densely arranged product of the last summer and autumn growth and the sharp contrast marks the periods of growth. To the two different regions of growth is given the term of "annual ring." By counting the number of these rings it is possible to estimate the age of the tree or branch.

Bark.—Bark or bork is a term applied to all that portion of a woody exogenous plant axis outside of the cambium line.

In pharmacognic work, bark is divided into three zones, these from without inward being:

1. OUTER BARK or CORK.

2. MIDDLE BARK or CORTICAL PARENCHYME.

3. INNER BARK or PHLOEM.

Commercially, bark is divided into two zones, which are, passing from without inward:

1 OUTER BARK (CORK).

2. INNER BARK (CORTICAL PARENCHYME AND PHLOEM).

Periderm.—Periderm is a name applied to all the tissue produced externally by the cork cambium (PHELLOGEN). This term appears often in pharmacognic and materia medica texts.

Histology of Typical Monocotyl Stems (Endogenous).—Passing from exterior toward centre the following structures are seen:

1. Epidermis whose cells are cutinized in their outer walls.

2. Hypodermis, generally collenchymatic.

3. Cortex.

4. Endodermis or innermost layer of cortex generally with greatly suberized cell walls.

5. A large central zone of parenchyme matrix in which are found scattered fibrovascular bundles of the closed collateral or rarely concentric type (amphivasal). In this latter type, which is typical of old monocotyl stems, the xylem grows completely around phloem so that phloem is found in the centre and xylem without and surrounding it. ·

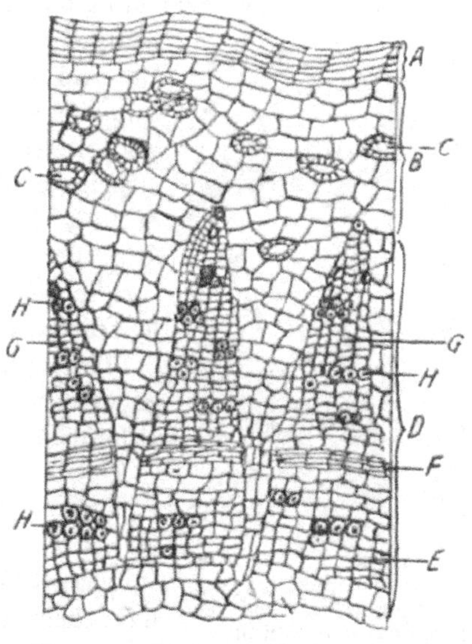

FIG. 23.—*Cinchona calisaya*. Cross-section of bark. *A*, Cork cells; *B*, cortical parenchyma; *C*, stone cells; *D*, phloëm porti n; *E*, soft bast; *F*, phellogen forming bark; *G*, medullary rays. (The black line from *G* should be extended to the parenchyma cells between the phloëm portions.) *H*, Bast fibers. (*From Sayre.*)

PLANT HAIRS OR TRICHOMES

These are out-growths of the epidermal cells which have become greatly elongated and may be unicellular or multicellular. They may be of various forms: *simple*, consisting of a single row of cells; branching; clavate, or club-shaped; stellate or star-shaped; barbed, hooked, forked, etc.

The terminal cell is often modified into a secretion sac for gummy, resinous or odorous products. Such hairs are called *glandular*. Ex.: Glandular hairs from strobiles of Humulus lupulus.

The cotton of commerce which is the hairs of the seed of the cotton plant, Gossypium herbaceum, is a good example of simple hairs.

Branched hairs can be seen upon the leaves of the common field weed, Mullein. Geranium and the Stinging nettle afford examples of glandular hairs.

Plant hairs are adapted to many different purposes. They absorb nourishment in the form of moisture and mineral matter in solution. Those which serve as a protection to the plant may be barbed and silicified, rendering them unfit for animal food, or, as in the nettle, charged with an irritating fluid, penetrating the skin when touched, injecting the poison into the wound. A dense covering of hairs also prevents the ravages of insects and the clogging of the stomata by an accumulation of dust. They fill an important office in the dispersion of seeds and fruits, as with their aid such seeds as those of the milkweed are readily scattered by the wind.

The reproductive organs of many Cryptogams are modified hairs, as the sporangia of Ferns.

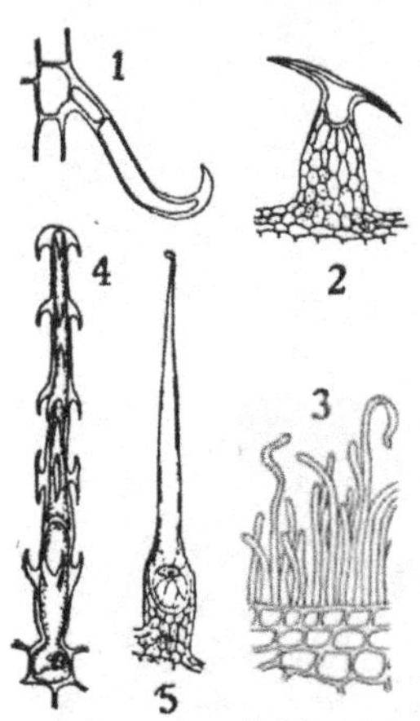

FIG. 24.—Different forms of epidermal outgrowths. 1, Hooked hair from Phaseolus multiflorus; 2, climbing hair from stem of Humulus Lupulus; 3, rodlike wax coating from the stem of Saccharum officinarum; 4, climbing hair of Loasa hispida; 5, stinging hair of Urtica urens. (*Fig. 3 after de Bary; the remainder from Häberlandt.*)

The Leaf

Leaves (folia) are stem appendages which have their origin just back of the apex of the stem, are regularly arranged upon it, and consist of expansions of its tissues.

The functions of a leaf are photosynthesis, assimilation, respiration and transpiration.

The most essential function of plants is the conversion of inorganic into organic matter; this takes place ordinarily in the green parts, containing chlorophyll, and in these when exposed to sunlight. Foliage is an adaptation for increasing the extent of green surface.

The leaf when complete consists of three parts, LAMINA, PETIOLE, and STIPULES. The lamina or blade is the expansion of the stem into a more or less delicate framework, made up of the branching vessels of the petiole.

The petiole is the leaf stalk. The stipules are leaf-like appendages appearing at the base of the petiole.

The leaf of the Tulip Poplar or Liriodendron affords a good example of a *Complete Leaf.*

Sometimes the lamina or blade is attached directly to the stem by its base and is then said to be *sessile.* If the petiole is present, *petiolate.*

When leaf stipules are absent, the leaf is said to be *exstipulate,* when present, *stipulate.*

The petiole is seldom cylindrical in form, but usually channelled

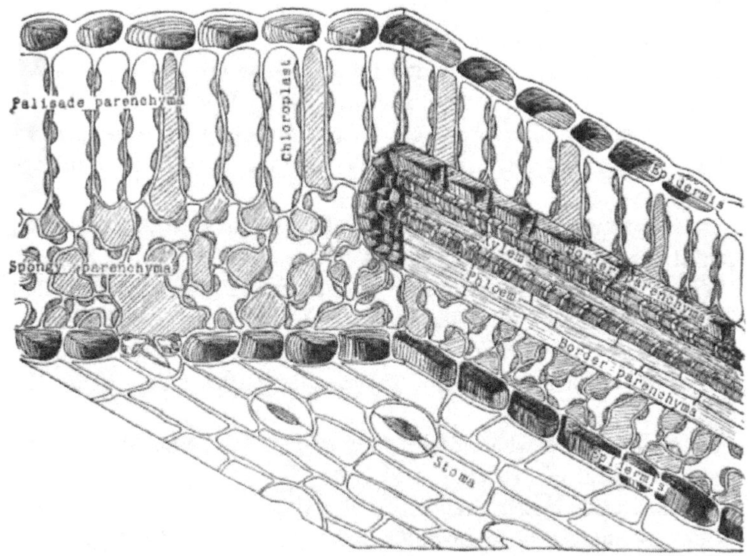

Fig. 25.—Stereogram of leaf structure. Part of a veinlet is shown on the right. Intercellular spaces are shaded. (*From Stevens.*)

on the upper side, flattened, or compressed. The stipules are always in pairs and closely resemble the leaf in structure.

The blade of the leaf consists of the framework, made up of branching vessels of the petiole, which are woody tubes pervading the soft tissue called mesophyll, or leaf parenchyme, and serve not only as supports but as veins to conduct nutritive fluids. Veins are absent in simple leaves such as many of the Mosses.

Leaf Venation.—FURCATE or FORKED VENATION is characteristic of many *Cryptogams.*

PARALLEL VENATION is typical of the *Monocotyledons*, as *Palms, Lilies, Grasses*, etc.

RETICULATE or NETTED VEINS characterize the *Dicotyledons*, as the *Poplar* or *Oak*.

PINNI-VEINED or FEATHER-VEINED leaves consist of a mid-vein with lateral veinlets extending from mid-vein to margin at frequent intervals and in a regular manner. Ex.: Calla.

PALMATELY VEINED leaves consist of a number of veins of nearly the same size, radiating from petiole to margin. Ex.: Maple leaf.

The Forms of Leaves.—SIMPLE LEAVES are those having a single blade, either sessile or petiolate.

COMPOUND LEAVES are divided into two or more distinct subdivisions called leaflets, which may be either sessile or petiolate.

Simple leaves and the separate blades of compound leaves are described as to general outline, apex, base, marginal indentations, surface and texture.

(*a*) GENERAL OUTLINE (form viewed as a whole without regard to indentations of margin). Dependent upon kind of venation.

When the lower veins are longer and larger than the others, the leaf is Ovate, or Egg-shaped. Parallel-veined leaves are usually linear, long and narrow of nearly equal breadth throughout, or lanceolate, like the linear with the exception that the broadest part is a little below the centre.

ELLIPTICAL, somewhat longer than wide, with rounded ends and sides. Ex.: Leaf of Pear.

OBLONG, when longer than broad, margins parallel. Ex.: Matico.

OBLIQUE, margin longer on one side than the other, as the *Hamamelis* and *Elm*.

ORBICULAR, circular in shape. Ex.: Nasturtium.

PELTATE, or shield-shaped, having the petiole inserted at the centre of the lamina. Ex.: the Nasturtium, Podophyllum.

FILIFORM, or THREAD-LIKE, very long and narrow, as *Asparagus* leaves.

OVATE, broadly elliptical. Ex.: Digitalis. *Obovate*, reversely ovate.

OBLANCEOLATE, reversely lanceolate. Ex.: Chimaphila.

CUNEATE, shaped like a wedge with the point backward.

SPATULATE, like a spatula, with narrow base and broad rounded apex. Ex.: Uva Ursi.

Acerose or Acicular, tipped with a needle-like point, as *Juniper*.

Deltoid, when the shape of the Greek letter Δ, as Chenopodium.

(*b*) **Apex of Leaf.**—Acute, when the margins form an acute angle at the tip of the leaf. Ex.: Eriodictyon.

Acuminate, when the point is longer and more tapering than the acute. Ex.: Pellitory.

Obtuse, blunt or round. Ex.: Buchu.

Truncate, abruptly obtuse, as if cut square off.

Mucronate, terminating in a short, soft point.

Cuspidate, like the last, except that the point is long and rigid.

Aristate, with the apex terminating in a bristle.

Emarginate, notched. Ex.: Pilocarpus.

Retuse, with a broad, shallow sinus at the apex.

Obcordate, inversely heart-shaped.

(*c*) **Base of Leaf.**—Cordate, heart-shaped. Ex.: Lime.

Reniform, kidney-shaped. Ex.: Ground Ivy.

Hastate, or halbert-shaped, when the lobes point outward from the petiole. Ex.: Aristolochia Serpentaria.

Auriculate, having ear-like appendage at the base.

Sagittate, arrow-shaped. Ex.: Bindweed.

(*d*) **Margin of Leaf.**—Entire, when the margin is an even line.

Serrate, with sharp teeth which incline forward like the teeth of a hand-saw. Ex.: Peppermint.

Dentate, or toothed, with outwardly projecting teeth. Chestnut.

Crenate, or Scalloped, similar to the preceding forms, but with the teeth much rounded. Ex.: Digitalis, Salvia.

Repand, or Undulate, margin—a wavy line.

Sinuate, when the margin is more distinctly sinuous than the last.

Incised, cut by sharp, irregular incisions. Ex.: Hawthorn.

Runcinate, the peculiar form of pinnately incised leaf observed in the Dandelion and some other Compositæ in which the teeth are recurved.

A Lobed leaf is one in which the indentations extend nearly to the mid-vein, or mid-rib, as it is usually called, the segments or sinuses, or both, being rounded. Ex.: Sassafras.

Cleft is the same as lobed, except that the sinuses are deeper, and commonly acute. Ex.: Dandelion.

A Parted leaf is one in which the incisions extend nearly to the mid-rib. Ex.: Geranium maculatum.

In the Divided leaf the incisions extend to the mid-rib, but the segments are not stalked. Ex.: Watercress.

If the venation is pinnate, the preceding forms may be described as pinnately incised, lobed, parted, or divided. If the venation is radiate, then the terms radiately or palmately lobed, incised, etc., are employed.

The transition from Simple to Compound Leaves is a very gradual one, so that in many instances it is difficult to determine whether a given form is to be regarded as simple or compound. The number and arrangement of the parts of a compound leaf correspond with the mode of venation, and the same descriptive terms are applied to outline, margin, etc., as in simple leaves.

Leaves are either pinnately or radiately compounded. They are said to be abruptly pinnate or paripinnate when the leaf is terminated by a pair of leaflets; odd pinnate or imparipinnate when it terminates with a single leaflet. When the leaflets are alternately large and small, the leaf is interruptedly pinnate, as the Potato leaf. When the terminal leaflet is the largest, and the remaining ones diminish in size toward the base the form is known as lyrate, illustrated in the leaf of the Turnip.

Radiately or palmately compound leaves have the leaflets attached to the apex of the petiole. When these are two in number the leaf is bifoliate, or binate; if three in number, trifoliate, or ternate; when four in number, quadrifoliate, etc. If each of the leaflets of a palmately compound leaf divides into three, the leaf is called bi-ternate; if this form again divides, a tri-ternate leaf results. Beyond this point the leaf is known as decompound. Example of decompound leaf seen in Cimicifuga.

Leaf Texture.—Leaves are described as:

Membranous, when thin and pliable, as Coca.

Succulent, when thick and fleshy, as Aloes, and Live Forever.

Scarious, when dry and scaly.

Coriaceous, when thick and leathery, as Eucalyptus, Uva Ursi and Magnolia.

Leaf Color.—Petaloid, when of some brilliant color different from the usual green, as the Coleus and Begonia, and other plants which are prized for the beauty of their foliage rather than their blossoms.

Leaf Surface.—Any plant surface is:

Glabrous, when perfectly smooth and free from hairs or protuberances.   Ex.: Tulip.

Glaucous, when covered with bloom, as the Cabbage leaf.

Pellucid-puncate, when dotted with oil glands, as the leaves of the Orange family.

Scabrous leaves have a rough surface with minute, hard points.

Pubescent, covered with short, soft hairs.   Ex.: Strawberry.

Villose, covered with long and shaggy hairs.   Ex. Forget-me-not.

Sericious, silky.   Ex.: Silverleaf.

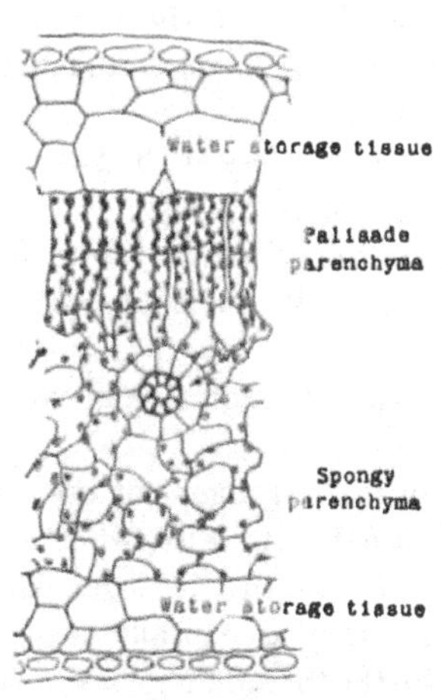

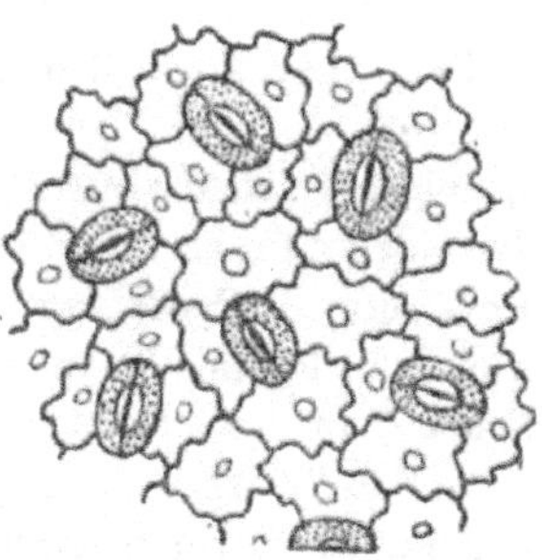

Fig. 26.   Fig. 27.

Fig. 26.—Cross-section through a portion of rubber leaf, showing the large percentage of water-storage tissues on both sides of the leaf, and the relation of the palisade and spongy parenchyma to the lateral veins.   (*From Stevens.*)

Fig. 27.—Surface view of the epidermis of a leaf showing several stomata.   The guard cells are dotted.   (*From Hamaker.*)

Hispid, when covered with short, stiff hairs.   Ex.: Borage.

Tomentose, densely pubescent and felt-like, as the Mullein leaf.

Spinose, beset with spines, as in the Thistle.

Rugose, when wrinkled.   Ex.: Sage.

Verrucose, covered with protuberances or warts.

Duration of Leaves.—Leaves vary as to their period of duration. They are: Persistent, or evergreen, if they remain green on the tree for a year or more.

Deciduous, if unfolding in spring and falling in autumn.

Caducous, or fugacious, if falling early in the season.

Like roots, they differ greatly as to duration in different latitudes. Evergreen trees are most common in the tropics, and it is probable that many of our deciduous trees have become such by adaptation to the colder climate.

Leaf Insertion.—The point of attachment of the leaf to the stem is called the insertion.   A leaf is:

RADICAL, when inserted upon an underground stem.

CAULINE, when upon an aërial stem.

RAMAL, when attached directly to a branch.

When the base of a sessile leaf is extended completely around the stem it is PERFOLIATE, the stem appearing to pass through the blade. Ex.: Urularia perfoliata or Mealy Bellwort.

When a sessile leaf surrounds the stem more or less at the base, it is called CLASPING.   Ex.: Poppy (Papaver somniferum).

When the bases of two opposite leaves are so united as to form one piece, they are called CONNATE-PERFOLIATE, as Eupatorium or Boneset.

Leaves are called EQUITANT when they are all radical and successively folded on each other, as the Iris.

Phyllotaxy.—Phyllotaxy is the study of leaf arrangement upon the stem or branch, and this may be either alternate, opposite, whorled, or verticillate, or fascicled.   It is a general law in the arrangement of leaves and of all other plant appendages that they are spirally disposed, or on a line which winds around the axis like the thread of a screw.   The spiral line is formed by the union of two motions, the circular and the longitudinal, and its most common modification is the circle.

In the ALTERNATE arrangement there is but one leaf produced at each node.

OPPOSITE, when a pair of leaves is developed at each node, on opposite sides of the stem.   Ex.: Mints, Lilac.

WHORLED or VERTICILLATE, when three or more form a circle about the stem.   Ex.: Canada Lily and Culver's root.

FASCICLED or TUFTED, when a cluster of leaves is borne from a single node, as in the Larch and Pine.

The spiral arrangement is said to be two-ranked when the third leaf is over the first, as in all Grasses; three-ranked, when the fourth is over the first. Ex.: Sedges.   The five-ranked arrangement is the most common, and in this the sixth leaf is directly over the first two turns being made around the stem to reach it.   Ex.: Cherry, Apple, Peach,

Oak and Willow, etc.   As the distance between any two leaves is two-fifths of the circumference of the stem, the five-ranked arrangement is expressed by the fraction 2/5.   In the eight-ranked arrangement the ninth leaf stands over the first, and three turns are required to reach it, hence the fraction 3/8 expresses it.   Of the series of fractions thus obtained, the numerator represents the number of turns to complete a cycle, or to reach the leaf which is directly over the first; the denominator, the number of perpendicular rows on the stem, or the number of leaves, counting along the spiral, from any one to the one directly above it.

Vernation.—PREFOLIATION or VERNATION relates to the way in which leaves are disposed in the bud.   A study of the individual leaf enables us to distinguish the following forms.   When the apex is bent inward toward the base, as in the leaf of the Tulip Tree, it is said to be INFLEXED or RECLINATE VERNATION; if doubled on the midrib so that the two sides are brought together as in the oak, it is CONDUPLICATE; when rolled inward from one margin to the other, as in the Wild Cherry, it is CONVOLUTE; when rolled from apex to base, as in Ferns, it is CIRCINATE; when folded or plaited, like a fan, it is PLICATE; if rolled inward from each margin, as the leaf of the common Violet, INVOLUTE; when rolled outward from each margin as Rumex, REVOLUTE.   The inner surface is always that which will form the upper surface when expanded.

Inflorescence or Anthotaxy.—A typical flower consists of four whorls of leaves modified for the purpose of reproduction, and compactly placed on a stem.   The term Inflorescence, Anthotaxy, is applied to the arrangement of the flowers and their position on the plant, both of which are governed by the same law which determines the arrangement of leaves. For this reason flower buds are always either terminal or axillary.   In either case the bud may develop a solitary flower or a compound inflorescence consisting of several flowers.

DETERMINATE, CYMOSE, DESCENDING, or CENTRIFUGAL inflorescence is that form in which the flower bud is terminal, and thus determines or completes the growth of the plant.   Ex.: Ricinus communis.

INDETERMINATE, ASCENDING, or CENTRIPETAL INFLORESCENCE is that form in which the flower buds are axillary, while the terminal bud continues to develop and increase the growth of the plant indefinitely. Ex.: the Geranium.

MIXED INFLORESCENCE is a combination of the other two forms. Ex.: Horse Chestnut.

4

The flower stalk is known as the PEDUNCLE, and its prolongation the RACHIS, or axis of the inflorescence.

The flower stalk of a single flower of an inflorescence is called a PEDICEL. When borne without such support the flower is SESSILE.

A peduncle rising from the ground is called a SCAPE, previously mentioned under the subject of stems.

The modified leaves found on peduncles are termed BRACTS. These vary much the same as leaf forms, are described in a similar manner, and may be either green or colored. When collected in a whorl at the base of the peduncle they form an INVOLUCRE, the parts of which are sometimes imbricated or overlapping, like shingles. This is generally

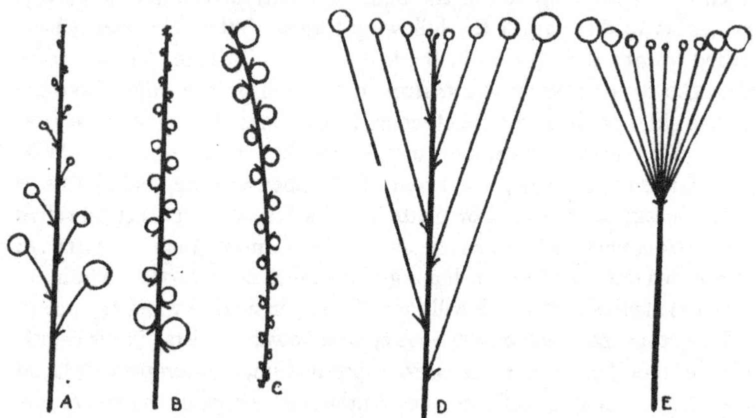

FIG. 28.—Types of racemose inflorescence; *A*, A raceme. *B*, A spike. *C*, A catkin. *D*, A corymb. *E*, An umbel. The flowers are represented by circles; the age of the flower is indicated by the size. (*From Hamaker.*)

green, but sometimes petaloid, as in the Dogwood. The modified leaves found on pedicels are called BRACTEOLAR LEAVES.

The SPATHE is a large bract enveloping the inflorescence and often colored, as in the Calla, or membranous, as in the Daffodil.

In the indeterminate or axillary anthotaxy, either flowers are produced from base to apex, those blossoming first which are lowest down on the rachis or from margin to centre. The principal forms of this type are: A solitary indeterminate is one in which the flowers occur singly in the axils of the leaves.

RACEME, or simple flower-cluster in which the flowers on pedicels of nearly equal length are arranged along an axis. Ex.: Convallaria, Cimicifuga, and Currant.

CORYMB, a short, broad cluster, differing from the raceme mainly in its shorter axis and longer lower pedicels, which give the cluster a flat appearance by bringing the individual florets to nearly the same level. Ex.: Cherry.

UMBEL, which resembles the raceme, but has a very short axis and the nearly equal pedicels radiate from it like the rays of an umbrella. Many examples of this mode of inflorescence are seen in the order *Umbelliferæ*, as indicated by the name, including Anise, Fennel and other official plants.

A SPIKE is a cluster of flowers, sessile or nearly so, borne on an elongated axis. The Mullein and common Plantain afford good illustrations.

The CATKIN or AMENT resembles the Spike, but differs in that it has scaly instead of herbaceous bracts, as the staminate flowers of the Oak, Hazel, Willow, etc.

The HEAD or CAPITULUM is like a spike, except that it has the rachis shortened so as to form a compact cluster of sessile flowers, as in the Dandelion, Marigold, Clover, and Burdock.

The STROBILE is a compact flower cluster with large scales concealing the flowers, as the inflorescence of the Hop.

The SPADIX is a thick, fleshy rachis with flowers closely sessile or embedded on it, usually with a spathe or sheathing bract. Ex.: Calla, Acorus, Calamus, Arum triphyllum.

The compound raceme particularly if irregularly compounded is called a PANICLE.

DETERMINATE ANTHOTAXY is one in which the first flower that opens is the terminal one on the axis, the others appearing in succession from apex to base or from centre to margin. The principal varieties are:

The SOLITARY DETERMINATE, in which there is a single flower borne on the scape, as in the Anemone, or Windflower, and Hydrastis.

The CYME, a flower cluster resembling a corymb, except that the buds develop from center to circumference. Ex.: Elder. If the cyme be rounded, as in the Snowball, it is a globose cyme.

A SCORPOID CYME imitates a raceme, having the flowers pedicelled and arranged along a lengthened axis.

A GLOMERULE is a cymose inflorescence of any sort which is condensed into a head, as the so-called head of Cornus florida.

A VERTICILLASTER is a compact, cymose flower cluster which resembles a whorl, but really consists of two glomerules situated in the axils of opposite leaves. Clusters of this kind are seen in Catnip, Horehound, Peppermint and other plants of the Labiatæ.

The raceme, corymb, umbel, etc., are frequently compounded. The compound raceme, or raceme with branched pedicels, is called a panicle. Ex.: Yucca and paniculate inflorescence of the oat.

A THYRSUS is a compact panicle, of a pyramidal or oblong shape. Ex.: Lilac, Grape and Rhusglabra.

A MIXED ANTHOTAXY is one in which the determinate and indeterminate plans are combined, and illustrations of this are of frequent occurrence.

The order of flower development is termed ascending when, as in

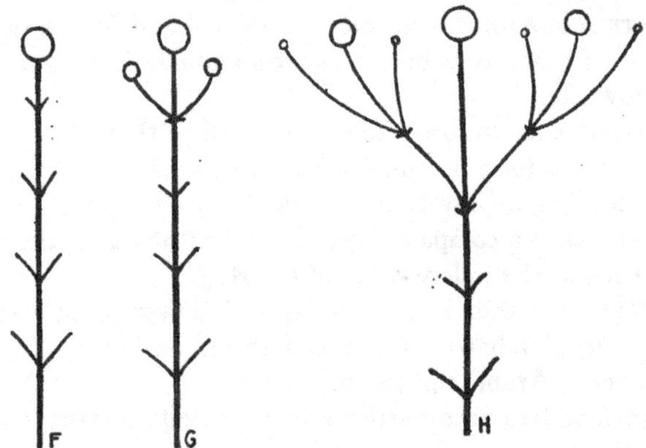

FIG. 29.—Cymose inflorescences. *F*, A terminal flower. *G*, A simple cyme. *H*, A compound cyme. (*From Hamaker.*)

the raceme, the blossoms open first at the lowest point on the axis and continue to the Apex. Ex.: White Lily, and many other plants of the same family. In the cyme the development is centrifugal, the central florets opening first, while in the corymb it is centripetal, or from margin to center.

Prefloration.—By prefloration is meant the arrangement of the floral envelopes in the bud. It is to the flower bud what vernation is to the leaf bud, the same descriptive terms being largely employed, as convolute, involute revolute, plicate, imbricate, etc.

In addition to those already defined, the following are important.

VALVATE PREFLORATION, in which the margins meet but do not overlap. Of this variety the induplicate has its two margins rolled inward as in Clematis. In the reduplicate they are turned outward, as the sepals of Althea.

VEXILLARY, the variety shown in the corolla of the Pea, where the two lower petals are overlapped by two lateral ones, and the four in turn overlapped by the larger upper ones.

CONTORTED, where one margin is invariably exterior and the other interior, giving the bud a twisted appearance, as in the Oleander and Phlox.

THE FLOWER

The flower is a shoot which has undergone a metamorphosis so as to serve as a means for the propagation of the individual.

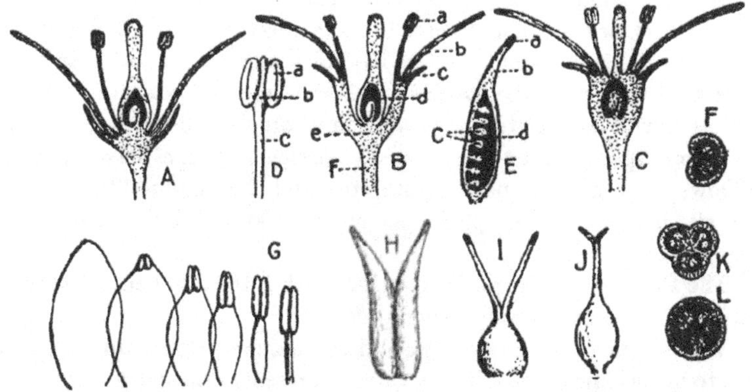

FIG. 30.—Diagrams of floral structures. *A*, shows the relations of the floral parts in a hypogynous flower. *B*, The same in a perigynous flower. *C*, The same in an epigynous flower. *D*, A stamen. *E*, A simple pistil in longitudinal section. *F*, The same in cross-section. *G*, Transitonal forms between true petals (left) and true stamens (right). *H*, Slight union of two carpels to form a compound pistil. *I* and *J*, Union of carpels more complete. *K* and *L*, Cross-sections of compound pistils, of three carpels. In *B: a*, stamen; *b*, petal; *c*, sepal; *d*, pistil; *e*, receptacle; *f*, pedicel. In *D: a*, anther cell; *b*, connective; *c*, filament. In *E: a*, stigma; *b*, style; *c*, ovules; *d*, ovary. (*From Hamaker.*)

The parts of the flower are SEPALS, PETALS, STAMENS, and CARPELS, all of which are inserted upon a shortened axis called the RECEPTACLE or TORUS. This is usually flat or convex, but may be conical and fleshy as in the Strawberry; concave as in the Rose and Fig; or show a disc-like modification, as in the Orange. The axis of a flower cluster, if short, is sometimes called a common receptacle, as in the floral axes of the Dandelion and Lettuce.

A *Complete Flower* possesses the four whorls of floral organs arranged upon the torus.

The stamens and pistils constitute the *essential* organs, and a flower is said to be *Perfect* when these are present and functional.

A *Regular Flower* possesses parts of the same shape and size.

It is *Symmetrical* when the parts of each whorl are of the same number, or multiples of the same number.

An *Imperfect Flower* shows one set of *essential* organs wanting.

When either petals or sepals, or both, are present in more than the usual number, the flower is said to be "*double*," as the cultivated Aster, Rose, and Carnation.

If the pistils are present and stamens wanting, the flower is called *pistillate*, or female; if it possesses stamens but no pistil, it is described as *staminate*, or male; if both are absent, *neutral*, as marginal flowers of Viburnum. Some plants, as the Begonias and Castor oil bear both staminate and pistillate flowers, and are called *Monœcious*. When the staminate and pistillate flowers are borne on different plants of the same species, they are termed *Diœcious*, as the Sassafras and Willow. When staminate, pistillate and hermaphrodite flowers are all borne on one plant, as on the Maple trees, they are *polygamous.*

The Perigone.—The perigone or perianth is the floral envelope consisting of calyx and corolla (when present).

When both whorls, *i.e.*, calyx and corolla, are present the flower is said to be dichlamydeous; if only calyx is present, monochlamydeous.

The Calyx.—The Calyx is the outer whorl of modified leaves. Its parts are called Sepals, and may be distinct (Chorisepalous, from a Greek word meaning disjoined) or more or less united (Gamosepalous). They are usually green—foliaceous or leaf-like—but may be brilliantly colored, hence the term petaloid (like the petals) is applied. Ex.: Tulip, Larkspur and Columbine.

In a GAMOSEPALOUS CALYX, when the union of sepals is incomplete, the united portion is called the tube, the free portion, the limb, the orifice of the tube, the throat.

In form the calyx may be regular or irregular; regular if its parts are evenly developed, and irregular if its parts differ in size and shape. The more common forms are tubular, resembling a tube; rotate, or wheel-shape; campanulate, or bell-shape; urceolate, or urn-shape; hypocrateriform, or salver-shape; bilabiate, or two-lipped; corresponding to the different forms of corolla, under which examples illustrating each will be given.

The calyx usually remains after the corolla and stamens have fallen, sometimes even until the fruit matures—in either case it is said to be persistent. If it falls with the corolla and stamens, it is deciduous, and if when the flower opens, caducous, as in the Poppy and May-apple. It is often more or less adherent to the ovary or base of the pistil, and it is important, in plant analysis, to note the presence or absence of such adhesion, which is indicated in a description by the terms inferior, or non-adherent (hypogynous), when free from the ovary and wholly beneath it; half-superior, or half-adherent (perigynous), when it partially envelops the ovary, as in the Cherry; superior or adherent (epigynous), when it completely envelops it, as in the Colocynth.

The Corolla.—The Corolla is the inner floral envelope, usually delicate in texture, and showing more or less brilliant colors and combinations of color. Its parts are called Petals, and when the calyx closely resembles the corolla in structure and coloring they are together called the Perianth. The purpose of these envelopes is to protect the reproductive organs within, and also to aid in the fertilization of the flower, as their bright colors, fragrance and saccharine secretions serve to attract pollen-carrying insects.

FORMS OF THE COROLLA AND PERIANTH.—When the petals are not united with each other, the corolla is said to be CHORIPETALOUS, often called Polypetalous. When more or less united, they are GAMOPETALOUS, often called SYMPETALOUS.

When the distinct petals are four in number, and arranged in the form of a cross, the corolla is called CRUCIFORM. Ex.: Mustard and other plants belonging to the order Cruciferæ.

The PAPILIONACEOUS COROLLA is so called because of a fancied resemblance to a butterfly. The irregularity in this form is very striking, and the petals bear special names: the largest one is the vexillum, or standard; the two beneath it the alæ, or wings; the two anterior, the carina or keel. Ex.: Locust, Pea, and Clover.

ORCHIDACEOUS flowers are of peculiar irregularity, combining calyx and corolla. The petal in front of stamen and stigma, which differs from the others in form and secretes nectar, is called the Labellum. Ex.: Cypripedium and other Orchids.

When calyx and corolla each consist of three parts closely resembling each other in form and color, as in the Tulip and Lily, the flower is called LILIACEOUS.

A GALEATE COROLLA is one in which the upper petal is arched in the shape of a helmet, called the Galea, as in Aconite.

The LIGULATE or STRAP-SHAPED COROLLA is nearly confined to the *family* Compositæ. It is usually tubular at the base, the remainder resembling a single petal. Ex.: Marigold, and Arnica Flowers.

LABIATE, or Bi-LABIATE, having two lips, the upper composed of two petals, the lower one of three. This form of corolla gives name to the LABIATÆ, while in the *family* LEGUMINOSÆ this arrangement is sometimes reversed. The corolla may be either ringent, or gaping, as in Sage, or personate, when the throat is nearly closed by a projection of the lower lip, as in Snapdragon.

ROTATE, WHEEL-SHAPED, when the tube is short and the division of the limb radiate from it like the spokes of a wheel. Ex.: The Potato blossom.

CRATERIFORM, SAUCER-SHAPED, like the last, except that the margin is turned upward or cupped. Ex.: Kalmia latifolia (Mt. Laurel).

HYPOCRATERIFORM, or SALVER-SHAPED (more correctly, hypocrateri-morphous), when the tube is long and slender, as in Phlox or Trailing Arbutus and abruptly expands into a flat limb. The name is derived from that of the ancient Salver, or hypocraterium with the stem or handle beneath.

When of nearly cylindrical form the corolla is TUBULAR, as in the Honeysuckle, and Stramonium.

FUNNEL-FORM (Infundibuliform), such as the corolla of the common Morning Glory, a tube gradually enlarging from the base upward into an expanded border or limb.

CAMPANULATE, or BELL-SHAPED, a tube whose length is not more than twice the breadth, and which expands gradually from base to apex. Ex.: Canterbury Bell, Harebell.

URCEOLATE, or URN-SHAPED, when the tube is globose in shape and the limb at right angles to its axis, as in the official Uva Ursi, Chimaphila and Gaultheria.

The Andrœcium, or Stamen System.—The Stamens or micro-sporophylls are the male organs of reproduction, and each complete stamen consists of a filament, or stalk, and an anther, or pollen sac, which is the essential portion and contains a powdery substance called pollen.

When few in number, stamens are said to be DEFINITE; when very numerous, and not readily counted, they are INDEFINITE. The following terms are in common use to express their number:

MONANDROUS, for a *flower* with but one stamen.

DIANDROUS, with two stamens.

TRIANDROUS, with three.

TETRANDROUS, with four.

PENTANDROUS, having five.

HEXANDROUS, six.

POLYANDROUS, an indefinite number.

As to insertion, they are:

HYPOGYNOUS, situated on the receptacle.

PERIGYNOUS, on the calyx tube or disc.

EPIGYNOUS, on the top of the ovary.

GYNANDROUS, borne upon the pistil, as in the Orchids.

The stamens may be of equal length; unequal, or of different length.

DIDYNAMOUS, when there are two pairs, one longer than the other.

TETRADYNAMOUS, three pairs, two of the same length, the third shorter.

Terms denoting connection between stamens are:

MONADELPHOUS (in one brotherhood), coalescence of the filaments into a tube.

DIADELPHOUS (in two brotherhoods), coalescence into two sets.

TRIADELPHOUS, with filaments united into three sets.

POLYADELPHOUS, when the sets are numerous.

SYNGENESIOUS, when the anthers cohere.

Stamens may be ERECT, extending directly upward, spreading, proceeding upward and outward; CONNIVENT brought close together and turned inward; REFLEXED, turned downward.

The attachment of the anther to the filament may be in one of several ways, as follows:

INNATE, attached at its base to the apex of the filament.

ADNATE, adherent throughout its length.

VERSATILE, when the anther is attached near its center to the top of the filament, so that it swings freely. The adnate and versatile are INTRORSE when they face inward, EXTRORSE when they face outward.

In order that the pollen may be discharged at the proper time, the anther opens along a line or suture called the line of dehiscence, either longitudinal or transverse, or the pollen escapes through apical or valvular openings. *The pollen* is usually a powdery substance which shows under the microscope distinct grains of characteristic forms, sizes and markings. Like starch grains, each represents a particular source, hence the variety that may be examined is limited only by the number of kinds of flowers available for the purpose. In order to study pollen grains, take up by means of forceps a stamen whose anther is just dehiscing, or letting free its contents, and tap upon a sheet of white paper; then examine with a Compound Microscope.

The following are some of the forms of pollen grains:

FOUR SPORE DAUGHTER cells hanging together as in the Cat Tail forming a pollen grain.

ELONGATED simple pollen grains as in Zostera.

DUMB-BELL shaped as the pollen of the Pines.

TRIANGULAR, as in the Mexican Primrose.

ECHINATE, as in the Malvaceæ.

SPHERICAL, as in Geranium.

LENS shaped as in the Lily.

The Gynœcium, or Pistil System.—The CARPEL or megasporophyll is the female organ of reproduction of flowering plants. In the Spruce, Pine, etc., it consists of an open leaf or scale which bears but does not enclose the *ovules*. In angiosperms it forms a closed sac which envelops and protects the *ovules*, and when complete is composed of three parts, the ovary or hollow portion at the base enclosing the *ovules* or rudimentary seeds, the *stigma* or apical portion which receives the pollen grains, and the *style*, or connective which unites these two. The last is non-essential and when wanting the stigma is called sessile. The carpel clearly shows its relations to the leaf, though greatly changed in form. The lower portion of a leaf, when folded lengthwise with the margins incurved, represents the *ovary*; the unfolded surface upon which the ovules are borne is the placenta, a prolongation of the tip of the leaf, the stigma, and the narrow intermediate portion, the style. *A leaf thus transformed into an ovule-bearing organ is called a carpel.* The carpels of the Columbine and Pea are made up of single carpels. In the latter the young peas occupy a double row along one of the sutures (seams) of the pod. This portion corresponds to the infolded edge of the leaf, and the pod splits open along this line, called the ventral suture.

Dehiscence, or the natural opening of the carpel to let free the contained seeds, takes place also along the line which corresponds to the mid-rib of the leaf, the dorsal suture.

COMPOUND PISTILS are composed of carpels which have united to form them, and therefore will have just as many cells as carpels. When each simple ovary has its placenta, or seed-bearing line, at the inner angle the resulting compound ovary has as many axile or central placentæ as there are carpels, but all more or less consolidated into one. The partitions are called dissepiments and form part of the walls of the ovary. If, however, the carpels are joined by their edges, like the petals of a gamopetalous corolla, there will be but one cell, and the placenta will be parietal, or on the wall.

The OVULES are transformed buds, destined to become seeds in the mature fruit. Their number varies from one to hundreds. In position, they are erect, growing upward from the base of the ovary, as in the Compositæ; ascending turning upward from the side of the ovary or cell; pendulous, like the last except that it turns downward; horizontal when directed straight outward; suspended, hanging perpendicularly from the top of the ovary.

In Gymnosperms the ovules are naked; in Angiosperms they are enclosed in a seed vessel.

A complete angiospermous seed ovule consists of a NUCELLUS or body; two coats, the outer or PRIMINE, and the inner or SECUNDINE; and a FUNICULUS, or stalk. Within the nucellus is found the EMBRYO SAC containing the OVUM or female reproductive cell.

The coats do not completely envelop the nucellus, but an opening at the apex, called the FORAMEN or MICROPYLE admits the pollen tube. The point where the coats are attached to each other and to the nucellus is called the CHALAZA. The HILUM marks the point where the funiculus is joined to the ovule, and if attached to the ovule through a part of its length, the adherent portion is called the RAPHE. The shape of the ovule may be ORTHOTROPOUS, or straight; CAMPYLOTROPOUS, bent or curved; AMPHITROPOUS, partly inverted; and ANATROPOUS, inverted. The last two forms are most common. A campylotropous ovule is one whose body is bent so that the hilum and micropyle are approximated.

THE PLACENTA

The placenta is the nutritive tissue connecting the ovules with the wall of the ovary. The various types of placenta arrangement (placentation) are grouped according to their relative complexity as follows:

1, Basilar.   2, Sutural.   3, Parietal.   4, Central.   5, Free Central.

Basilar placentation is well illustrated in the Polygonaceæ (Smart Weed, Rhubarb, Etc.) in Piper and Juglans.   Here at the apex of the axis and in the center of the ovarian base arises a single ovule from a small area of placental tissue.

Sutural placentation is seen in the Leguminosæ (Pea, Bean, Etc.). Here each carpel has prolonged along its fused edges two cord-like placental twigs, from which start the funiculi or ovule stalks.

Parietal placentation is seen in Gloxinia, Gesneria, Etc.   Here we find two or more carpels joined and placental tissue running up along edges of the fused carpels bearing the ovules.

Central or axile placentation is seen in Campanulaceæ (Lobelia), where the two, three, or more carpels have folded inward until they meet in the center and in the process have carried the originally parietal placenta with them.   This then may form a central swelling bearing the ovules over the surface.

Free Central placentation occurs perfectly in the Primulaceæ, Plantaginaceæ and a few other families.   In this the carples simply cover over or roof in a central placental pillar around which the ovules are scattered.

Pollination.—Pollination is the transfer of pollen from anther to stigma and the consequent germination thereon.   It is a necessary step to fertilization.

When the pollen is transferred to the stigma of its own flower the process is called CLOSE or SELF POLLINATION; if to a stigma of another flower, CROSS POLLINATION.   If fertilization follows, these processes are termed respectively, Close or Self Fertilization and Cross Fertilization.   Close Fertilization means in time ruination to the race and happily is prevented in many cases by (a) the stamens and pistils standing in extraordinary relation to each other, (b) by the anthers and pistils maturing at different times, (c) by the pollen in many cases germinating better on the stigma of another flower than its own.

The agents which are responsible for cross pollination are the wind, insects, water currents, small animals, and birds.

WIND-POLLINATED flowering plants are called ANEMOPHILOUS; their pollen is dry and powdery, flowers inconspicuous and inodorous, as in the Pines, Wheat, Walnut, Hop, etc.

INSECT-POLLINATED PLANTS are called ENTOMOPHILOUS.   These, being dependent upon the visits of insects for fertilization, possess brilliantly colored corollas, have fragrant odors, and secrete nectar, a sweet liquid very attractive to insects which are adapted to this work through the possession of a pollen-carrying apparatus.   Ex.: Orchids.

PLANTS POLLINATED THROUGH THE AGENCY OF WATER CURRENTS are known as HYDROPHILOUS.   To this class belong such plants as live under water and which produce flowers at or near the surface of the same.   Ex.: Sparganium.

Some plants as the Honeysuckle and Nasturtium are fertilized by humming birds.

Before the pollen grain has been deposited upon the stigma a series of events affecting both the pollen grain and the embryo sac occur. The microspore (pollen grain) divides into two cells, the MOTHER and TUBE CELLS of the male gametophyte.   The nucleus of the mother cell divides to form two generative nuclei.   The NUCLEUS of the mega-spore or embryo sac undergoes division until eight DAUGHTER NUCLEI are produced which are separated into the following groups:

(a) Three of these nuclei occupy a position at the apex, the lower nucleus of the group being the egg or ovum, the other two nuclei being the SYNERGIDS or ASSISTING NUCLEI.

(b) At the opposite end of the sac are three nuclei known as the ANTIPODALS which apparently have no special function.

(c) The two remaining nuclei (POLAR NUCLEI) form a group lying near the centre of the embryo sac which unite to form a single nucleus from which, after fertilization, the endosperm of nourishing material is derived.   This stage of the embryo sac constitutes the female gametophyte.

Fertilization.—After the pollen grain reaches the stigma the viscid moisture of the stigma excites the outgrowth of the male gametophyte which bursts through the coats of the pollen grain forming a pollen tube.   The pollen tube carrying within its walls two generative and one tube nucleus penetrates through the loose cells of the style until it reaches the micropyle of the ovule, then piercing the nucellus it enters the embryo sac.   The tip of the tube breaks and one of the generative nuclei unites with the egg to form the oöspore.   The oöspore develops at once into an embryo or plantlet, which lies passive until the seed undergoes germination.   The other generative nucleus unites with the previously fused polar nuclei to form the endosperm nucleus which soon undergoes rapid division into a large number of nuclei scattered about through the protoplasm of the embryo sac.   These accumulate protoplasm about them, cells walls are laid down, endosperm resulting.

Germination is the beginning of growth in a seed or plant. The conditions favorable to germination are warmth, moisture and presence of air.

The Fruit

The fruit consists of the matured ovary and contents, and may include other organs of the flower external to the pistil, but connected with it, as in Clematis, where the long, feathery style renders the fruit buoyant, and, like the fruits of the Thistle and Dandelion, in which the modified calyx serves a similar purpose, is easily scattered by the wind. In the Strawberry and Quince the receptacle becomes thick and succulent, and constitutes the edible portion of the fruit. Other modifications are seen in the hooks or spines, by means of which certain fruits compel animals to assist in their dispersion. Ex.: Cocklebur, Burdock, Bidens, Etc.

Distribution of Fruits and Seeds.—Some fruits, as the cocoanut, are transported by water currents, and are adapted to withstand for a long period the action of salt water. Another of the peculiar means provided by nature for the dissemination of seeds and fruits is that shown in the Sandbox Tree, the fruit of which is hygroscopic, and by absorption of water bursts the pericarp with such explosive force as to cause a loud report and to scatter the seeds in every direction. Birds and fruit-eating mammals, including man, also play a part in the work of distribution.

Fruit Structure

The PERICARP, or SEED VESSEL, is the ripened wall of the ovary, and in general the structure of the fruit wall resembles that of the ovary, but undergoes numerous modifications in the course of development.

The number of cells of the ovary may increase or decrease, the external surface may change from soft and hairy in the flower to hard, and become covered with sharp, stiff prickles, as in the Datura Stramonium or Jamestown weed. Transformations in consistence may take place and the texture of the wall of the ovary may become hard and bony, leathery, as the rind of the Orange, or assume the forms seen in the Gourd, Peach, Grape, etc.

Where the pericarp consists of two layers of different texture, as in the Plum, the outer layer is called EXOCARP, the inner, ENDOCARP. When the external layer is thin, it is sometimes termed the EPICARP, when the middle or inner layers are fleshy or pulpy they constitute the SARCOCARP.

When the endocarp within the sarcocarp is hard, forming a shell or stone, this is termed a PUTAMEN. When three concentric layers are distinguishable in a pericarp, the middle one is called mesocarp.

Fruits are either DEHISCENT or INDEHISCENT according as they discharge or retain their seeds. Dehiscent fruits open regularly, or normally. When the pericarp splits vertically through the whole or a part of its length, along sutures or lines of coalescence of contiguous carpels. Legumes usually dehisce by both sutures. Irregular or abnormal dehiscence has no reference to normal sutures, as where it is transverse or circumscissile, extending around the capsule either entirely or forming a hinged lid, as in Hyoscyamus, or detached.

DEHISCENCE is called POROUS or APICAL when the seeds escape through pores at the apex, as in the Poppy; valvular, when valve-like orifices form in the wall of the capsule. VALVULAR DEHISCENCE is SEPTICIDAL when the constituent carpels of a pericarp become disjoined, and then open along their ventral suture; LOCULICIDAL, DEHISCENCE into loculi, or cells, in which each component carpel splits down its dorsal suture, and the dissepiments remain intact; SEPTIFRAGAL DEHISCENCE, a breaking away of the valves from the septa or partitions.

Classification of Fruits (according to structure).—SIMPLE FRUITS result from the ripening of a single pistil in a flower.

AGGREGATE FRUITS are the product of several distinct pistil ripenings in one flower, the cluster of carpels being crowded on the receptacle ing one mass, as in the Raspberry, Blackberry, and Strawberry.

MULTIPLE FRUITS are those which are the product of a flower cluster instead of a single flower.

Simple Fruits are either Dry or Fleshy. The first may be divided into Dehiscent, those which split open when ripe; and Indehiscent, those which do not.

Dry, Dehiscent, Simple and Aggregate Fruits.—The FOLLICLE is a pod formed by a simple pistil and dehiscent by one suture, as Aconite and Staphisagria.

A LEGUME is a pod formed by a simple pistil and dehiscent by both sutures. The name legume is restricted to the fruits of the natural family Leguminosæ, and includes all the modifications which it represents.

A jointed, indehiscent legume, called a loment, breaks up naturally into transverse, one-seeded divisions. The Cochlea is a coiled or spiral legume. Ex. of Loment: Cassia fistula.

A CAPSULE is a dry dehiscent fruit of two or more united carpels, and shows several forms of dehiscence, as in the Poppy, Cardamon, etc.

The PYXIS is a modification of the capsule which opens transversely, the upper half forming a lid, as in Portulaca or Hyoscyamus.

A SILIQUE is a long slender capsule with two parietal placentæ, the valves opening from below upward, as in the Cruciferæ.

Dry Indehiscent Fruits (often erroneously regarded as seeds).-- The AKENE is a dry one-chambered, indehiscent fruit, in which the pericarp is firm and may or may not be united with the seed, the style remaining in many cases as an agent of dissemination, and may be winged, feathery, or hooked. Ex.: Fruits of the Compositæ, Anemone and Ranunculus.

The SAMARA is a winged akene-like fruit, as in the Birch, Elm, Ash, Box Elder and Maple.

The UTRICLE is like the akene, except that the pericarp is loose and bladder-like. Ex.: Chenopodium.

A CARYOPSIS, or Grain, differs from the last in having the cell completely filled by the seed and the pericarp very thin. This fruit is more likely than any other to be mistaken for a seed. Ex.: Wheat, Rice, Barley, Oat, etc.

A NUT is a hard, one-celled, one-seeded fruit, like the akene but larger, and usually produced from a compound ovary. The nut is often enclosed in a kind of involucre termed a Cupule, as the cup of the acorn or the leaf-like covering of the Hazel-nut.

A CREMOCARP is the characteristic fruit of the Umbelliferæ family. It consists of two inferior akenes or mericarps separated from each other by a stalk called a carpophore. The mericarps separate as soon as the fruit ripens and are seen to be longitudinally ribbed with numerous oil glands between the ribs.

Fleshy Indehiscent Fruits.—The DRUPE is a one-carpelled fruit, such as the Plum, Peach, Prune, Sabal, Rhus, etc., and called "stone fruit," because the endocarp or putamen is composed wholly of stone cells.

An ETÆRIO consists of a collection of little drupes on a torus as the Raspberry.

The BERRY is fleshy fruit with a thin membranous epicarp and a succulent interior in which the seeds are imbedded. Ex.: Capsicum, Tomato, Belladonna, Grape, Currant, etc.

The HESPERIDIUM is a variety of the berry, and the name is applied only to members of the Orange family. It is a fleshy fruit with leathery rind which contains numerous oil glands.

The PEPO or GOURD-FRUIT, of which the Squash and Gourd are types, is the characteristic fruit of the order Cucurbitaceæ, fleshy internally, and having a tough or very hard rind. Fruits of this family are true berries.

The POME is a fleshy fruit the chief bulk of which consists of the adherent torus. Quince, Apple and Pear are examples. The carpels constitute the core, and the fleshy part is developed from the torus.

Multiple Fruits.—The SYCONIUM is a multiple fruit consisting of a succulent hollow torus enclosed within which are akene-like bodies, products of many flowers. Ex.: Fig.

The SOROSIS is represented by the Mulberry, the grains of which are not the ovaries of a single flower, as in the Blackberry, but belong to as many separate flowers. In the Pine-apple all the parts are blended into a fleshy, juicy, seedless mass, and the plant is propagated by cuttings.

The STROBILE or CONE is a scaly, multiple fruit consisting of a scale-bearing axis, each scale enclosing one or more seeds. The name is applied to the fruit of the Hop, and also to the fruit of the Coniferæ in which the naked seeds are borne on the upper surface of the woody scales.

THE SEED

The seed is the fertilized and matured ovule, having the embryo formed within it. Like the ovule, it consists of a nucellus or kernel enclosed by integuments, and the descriptive terms used are the same. The seed coats, corresponding to those of the ovule, are two in number, the TESTA and TEGMEN. The testa, or outer seed shell, differs greatly in form and texture. If thick and hard, it is crustaceous; if smooth and glossy, it is polished; if roughened, it may be pitted, furrowed, hairy, reticulate, etc.

The testa may often present outgrowths or seed appendages whose functions are to make the seeds buoyant, whereby they may be disseminated by wind currents. Examples of these are seen in the Milk-weed, which has a tuft of hairs at one end of the seed called a *Coma*, and in the official Strophanthus, which has a long bristle-like appendage attached to one end of the seed and called an *awn*. The wart-like appendage at the hilum or micropyle, as in Castor Oil Seed, is called the Caruncle.

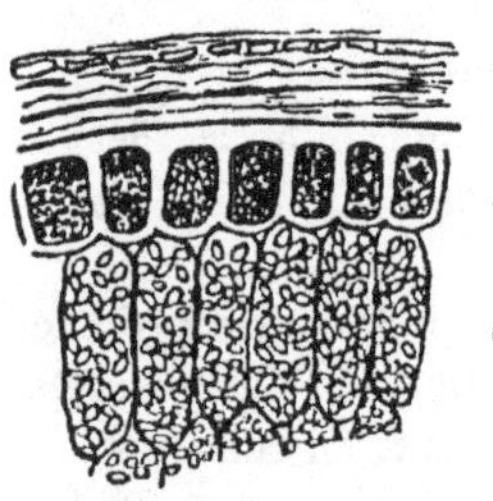

FIG. 31.—Section of a grain of wheat. *A*, Pericarps and seed coats; *B*, layer of cells in endosperm containing aleurone grains; *C*, cells of the endosperm containing starch grains. (*From Hamaker.*)

The tegmen or inner coat surrounds the nucellus closely and is generally soft and delicate.

A third integument, or accessory seed covering, is occasionally present and is called the ARIL. Ex.: Euonymus (succulent).

When such an integument arises at the micropyle of the seed, as in the Nutmeg, it is known as an arillode.

The NUCELLUS or KERNEL consists of tissue containing albumen, when this substance is present, and the embryo. Albumen is the *name* given the nutritive matter stored in the seed.

MODE OF FORMATION OF DIFFERENT TYPES OF ALBUMEN

If the egg cell within the embryo sac segments and grows into the embryo and, stretching, fills up the cavity without food material laid down around it, it happens that the nutritive material lingers in the cells of the nucellus pressing around the embryo. This is called *Perispermic Albumen.* Seen in the Polygonaceæ.

In by far the greater number of Angiosperms, the endosperm nucleus, after double fertilization, divides and redivides, giving rise to numerous nuclei imbedded in the protoplasm of the embryo sac outside of the developing embryo. Gathering protoplasm about themselves and laying down cell walls they form the endosperm tissue outside of the embryo. Into this tissue food is passed constituting the Endospermic albumen.

In the Marantaceæ, Piperaceæ, etc., nutritive material is passed into the nucellar cells causing them to swell up, while to one side a small patch of endosperm tissue accommodates a moderate amount of nourishing substance, thus resulting in the formation of abundant perisperm and a small reduced endosperm.

Ezalbuminous seeds are those in which the albumen is stored in the embryo during the growth of he seed.

Albuminous seeds are those in which the nourishment is not stored in the embryo until germination takes place.

PART II

TAXONOMY

DIVISION I.—THALLOPHYTA

Plants consisting of a thallus, a body undifferentiated into root, stem or leaf. The group nearest to the beginning of the plant kingdom presenting forms showing rudimentary structures which are modified through division of labor, differentiation, etc., in higher groups.

SUBDIVISION I.—MYXOMYCETES, OR SLIME MOLDS

Terrestrial or aquatic organisms, frequently classified as belonging to the animal kingdom and found commonly on decaying wood, leaves, or humous soil in forests. Their vegetative body consists of a naked mass of protoplasm called the plasmodium which has a creeping and rolling motion, putting out and retracting regions of its body called pseudopodia.

SUBDIVISION II.—SCHIZOPHYTA

This group comprises the "fission plants" whose members possess a common method of asexual reproduction whereby the cell cleaves or splits into two parts, each of which then becomes a separate and independent organism.

1. CYANOPHYCEÆ

Plants which are sometimes termed blue-green algæ. They contain chlorophyll, a green pigment and phycocyanin, a blue pigment, a combination giving a blue-green aspect to the plants of this group. Found everywhere in fresh and salt water and also on damp logs, rocks, bark of trees, stone walls, etc. Ex.: Oscillatoria, Glœocapsa, and Nostoc.

2. SCHIZOMYCETES—BACTERIA

Bacteria are minute, unicellular vegetable organisms destitute of chlorophyll. They serve as agents of decay and fermentation and are frequently employed in industrial processes. According to the various phenomena they produce they may be classified as follows: a. Zymogens producing fermentation; b. Aerogens producing gas; c. Photogens producing light; d. Chromogens producing color; e. Saprogens, producing putrefaction; f. Pathogens, producing disease.

PHYSICAL APPEARANCE OF BACTERIAL COLONIES AND INDIVIDUAL FORMS

Because of their minute size—a space the size of a pinhead may hold 8 billion of them—the student commences his study of bacterial growths in colonies or cultures, each kind possessing characteristics by which they may be distinguished and differentiated.

The individuals in the colony, depending upon the kind of bacteria under examination, may be globular, rod-shaped, or spiral. Bacteria are classed according to shape, as

COCCI (singular coccus), globular or berry-shaped.

BACILLI (singular, bacillus—a little rod), rod-shaped.

SPIRILLA (singular, spirillum), spiral or corkscrew-shaped.

Sporulation.—A large number of bacteria possess the power of developing into a resting stage by a process known as sporulation or spore formation. Sporulation is regarded as a method of resisting unfavorable environment. This is illustrated by the anthrax bacilli which are readily killed in twenty minutes by a 10 per cent. solution of carbolic acid, and able, when in the spore condition, to resist the same disinfectant for a long period in a concentration of 50 per cent. And, while the vegetative forms show little more resistance against moist heat than the vegetative form of other bacteria, the spores will withstand the action of live steam for as long as ten to twelve minutes or more.

Whenever the spores are brought into favorable condition for bacterial growth, as to temperature, moisture and nutrition, they return to the vegetative form and then are capable of multiplication by fission in the ordinary way.

Reproduction.—Bacteria multiply and reproduce themselves by cleavage or fission. A young individual increases in size up to the limits of the adult form, when by simple cleavage at right angles to the long axis, the cell divides into two individuals.

Morphology Due to Cleavage.—According to limitations imposed by cleavage directors, the cocci assume a chain appearance, or a grape-like appearance, or an arrangement in packets or cubes having three diameters. This gives rise to the

STAPHYLOCOCCUS (plural, staphylococci), from a Greek word referring to the shape of a bunch of grapes.

STREPTOCOCCUS (plural, streptococci), from a Greek word meaning chain-shaped.

SARCINA, package shaped or cubical.

Form of Cell Groups after Cleavage.—The individual bacteria after cleavage may separate, or cohere. The amount of cohesion, together with the plane of cleavage, determines the various forms of the cell groups. Thus among the cocci diplo- or double forms may result giving rise to distinguishing morphological characteristics. Similarly among the bacilli characteristic forms result as single individuals and others which form chains of various lengths.

Rapidity of Growth and Multiplication.—The rapidity with which bacteria grow and multiply is dependent upon species and environment. The rapidity of the growth is surprising. Under favorable conditions they may elongate and divide every 20 or 30 minutes. If they should continue to reproduce at this rate for twenty-four hours a single individual would have 17 million descendants. If each of these should continue to grow at the same rate, each would have in twenty-four hours more, 17 million offspring, and then the numbers would develop beyond conception. However, such multiplication is not possible under natural or even artificial conditions, both on account of lack of nutritive material and because of the inhibition of the growth of the bacteria by their own products. If they did multiply at this rate in a few days there would be no room in the world but bacteria.

Chemical Composition of Bacteria.—The quantitative chemical composition of bacteria is subject to wide variations, dependent upon the nutritive materials furnished them. About 80 to 85 per cent. of the bacterial body is water; proteid substances constitute about 50 to 80 per cent. of the dry residue. When these are extracted, there remain fats, in some cases wax, in some bacteria traces of cellulose appear, and the remainder consists of 1 to 2 per cent. ash.

The proteids consist partly of nucleo-proteids, globulins, and protein substances differing materially from ordinary proteids. Toxic substances known as endotoxins to distinguish them from bacterial poisons secreted by certain bacteria during the process of growth, also occur.

SUBDIVISION III.—ALGÆ

Low forms of thallophytes of terrestrial and aquatic distribution consisting for the most part of single cells or rows of single cells joined end to end to form filaments. They contain chlorophyll or some other pigment, and so can use the CO_2 and H_2O in the same manner as higher plants, *e.g.*, in assimilating and providing for their own nutrition.

CLASS I.—CHLOROPHYCEÆ, THE GREEN ALGÆ

In this group the cells are observed to possess distinct nuclei and bodies, whose pores contain an oil-like pigment called chlorophyll, the chloroplasts. The following forms are typical: Spirogyra, Diatoms, Pleurococcus, Volvox, Conferva, and Chara.

CLASS II.—PHÆOPHYCEÆ, THE BROWN ALGÆ

Mostly marine forms showing great diversity in the form of their vegetative bodies. Their bodies are usually fixed to some support in the water and are often highly differentiated both as to form and tissues. Some reach hundreds of feet in length as, for example, Macrocystis which grows in the Pacific Ocean off the coast of California. Other forms typical of the group are Ectocarpus, Laminaria, and Fucus.

CLASS III.—RHODOPHYCEÆ, THE RED ALGÆ

A greatly diversified group comprising the majority of marine algæ. Their vegetative bodies vary from simple branching filaments through all gradations to forms differentiated into branching stems, holdfasts and leaves. Their color may be red, purple, violet, or reddish brown and is due to the presence of phycoerythrin, a red pigment. Among this group are classed Chondrus, Nemalion, Corallina, etc. Chondrus is the sole official alga in the U. S. P. and belongs to the family Gigartinaceæ.

SUBDIVISION IV.—FUNGI

This great assemblage of thallophytes is characterized by the total absence of chlorophyll and so its members possess no independent power of manufacturing food materials such as starches, sugars, etc., from CO_2 and H_2O. Consequently they are either parasites, depending for their nourishment upon other living plants or animals, called hosts; or saprophytes, depending upon decaying animal or vegetable matter in solution. Some forms are able to live either as saprophytes or parasites while others are restricted to either the parasitic or saprophytic habit. The vegetative body of a fungus is known as a mycelium. It consists of interlacing and branching filaments called hyphæ, which ramify through decaying matter or invade the tissues of living organisms and derive nourishment therefrom. In the case of parasites, the absorbing connections which are more or less specialized and definite are called haustoria. In the higher forms the hyphæ become consolidated into false tissues, and assume definite shapes according to the species. Of this character are the fructifying organs which constitute the above ground parts of Puff Balls, Cup Fungi, Mushrooms, etc.

CLASS I.—PHYCOMYCETES, OR ALGA-LIKE FUNGI

The Phycomycetes represent a small group of fungi showing close affinity with the green algæ. Their mycelium is composed of cœnocytic hyphæ which suggests a close relation with the Siphonales group of green algæ. Their sexual organs are likewise similar in structure.

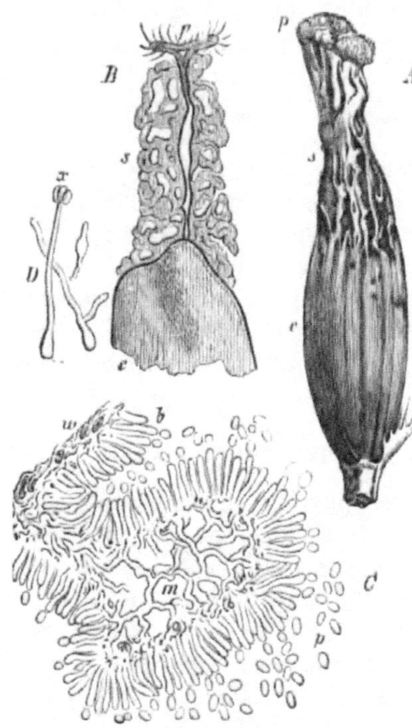

FIG. 32.—*Clasiceps purpurea.* A. Young sclerotium, *s*, with old sphacelia, *s.p.* The apex of the dead ovary of rye. B. Upper part of *A*, in longitudinal section, showing sphacelia, *s*. C. Transverse section through the sphacelia, more highly magnified. *m*. The mycelium, surrounded with the hyphæ; *b*, bearing conidia; *p*. conidia fallen off; *w*, the wall of the ovary. D. Germinating conidia, forming sporidia,*x*. (*From Sayre after Bachs.*)

FIG. 33.—Portion of Horn-shaped sclerotium of *Clasiceps purpurea*, bearing four stalked receptacles. (*From Sayre.*)

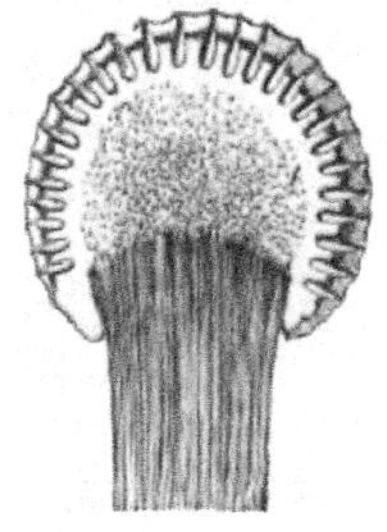

FIG. 34.—Longitudinal section of a receptacle, magnified, showing the perithecia. (*From Sayre.*)

SUB-CLASS A.—OöMYCETES

(Sexual apparatus heterogamous)

Order 1. Chytridiales.—Ex.: Synchytrium, a form parasitic on seed plants and forming blister-like swellings.

Order 2. Saprolegniales.—Water molds which attack fishes, frogs, water insects, and decaying plants and animals. Ex.: Saprolegnia.

Order 3. Peronosporales.—Mildews, destructive parasites, living in the tissues of their hosts and effecting pathologic changes. Ex.: Albugo, the blister blight, a white rust attacking members of the Cruciferæ and Phytophthora, producing potato rot.

SUB-CLASS B.—ZYGOMYCETES

(Sexual apparatus shows isogamy)

Order 1. Mucorales, the black molds, mostly saprophytic. Ex.: Mucor Mucedo, Rhizopus nigricans.

CLASS II.—ASCOMYCETES, THE SAC FUNGI

Mycelium composed of septate filaments and life history character-

FIG. 35.—A single perithecium of *Claviceps purpurea*, magnified, showing the contained asci. (*From Sayre.*)

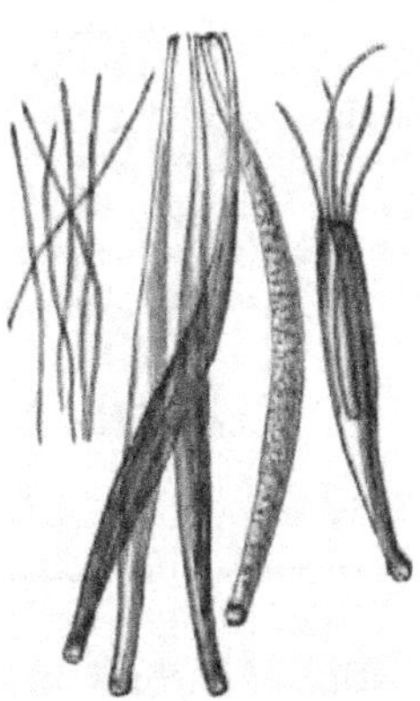

FIG. 36.—Asci containing the long, slender ascospores. (*From Sayre.*)

ized by the appearance of a sac called an ascus in which ascospores are formed. The largest class of fungi.

Order 1. Protoascales, the yeasts (Saccharomycetes) (position doubtful).

Order 2. Pezizales or cup fungi. Ex.: Peziza and Ascobolus.

Order 3. Tuberales, the truffles.

Order 4. Plectascales, the blue and green molds. Ex.: Aspergillus and Penicillium.

Order 5. Pyrenomycetales, the mildews and black fungi common as superficial parasites on various parts of plants. To the black fungi division of this order the Ergot fungus, Claviceps purpurea, belongs.

LIFE HISTORY OF CLAVICEPS PURPUREA

Through the agency of winds or insects the spores (ascospores or conidia) of this organism are brought to the young ovaries of the rye (Secale cereale). They germinate into long filaments called hyphæ which, becoming entangled to form a mycelium, spread over the ovary, enter it superficially, secrete a ferment, and cause decomposition of its tissue and the resultant formation of a yellow-mucus substance called honey-dew, which surrounds chains of moniliform reproductive bodies known as conidia. The honey-dew attracts certain insects which disseminate the disease to other heads of grain.

The mycelial threads penetrate deeper and deeper into the ovary and soon form a dense tissue which gradually consumes the entire substance of the ovary and hardens into a purple somewhat curved body called a sclerotium, or official ergot—the resting stage of the fungus, Claviceps.

The ergot falls to the ground and in the following spring sprouts into several stalked heads. Each (fruiting) head or ascocarp has imbedded in its surface numerous flask-shaped invaginations called perithecia from the bases of which several sacs or asci develop. Within each ascus are developed eight filiform spores (ascospores) which, when the ascus ruptures, are discharged and are carried by the wind to other fields of grain, there to begin over a new life cycle.

CLASS III.—BASIDIOMYCETES, OR BASIDIA FUNGI

This large class of fungi including the smuts, rusts, mushrooms, gill and tooth fungi, etc., is characterized by the occurrence of a basidium in the life history. A basidium is the swollen end of a hypha consisting of one or four cells and giving rise to branches called sterigmata, each of which cuts off at its tip a spore.

SUB-CLASS A.—PROTOBASIDIOMYCETES

(Basidium four-celled, each cell bearing a spore)

Order 1. Ustilaginales, the smuts. Destructive parasites which attack the flowers of various cereals, occasionally other parts of these plants. Ex.: Ustilago Maydis, the corn smut.

Order 2. Uredinales, the rusts. Ex.: Puccinia graminis, one of the wheat rusts, living in the intercellular spaces of young wheat.

(Basidium one-celled within which either four spores or four nuclei are formed)

Division a.—Hymenomycetes

(Hymenium or spore-bearing surface exposed)

Order 1. Thelephorales, forms appearing on tree trunks as leathery incrustations or as bracts on the ground, old logs, etc.

Order 2. Clavariales, the coral fungi. Fleshy coral or club-shaped forms. Ex.: Clavaria.

Order 3. Agaricales, the mushroom or toadstool alliance. Alike with the other members of the Basidiomycetes the plant body consists of the mycelium, ramifying through the substratum, but the part which rises above the surface (the Sporophore) is differentiated into a stalk-like body called a stipe bearing upon its summit a cap or pileus, the latter having special surfaces for the hymenium.

Family 1. Hydnaceæ or tooth fungi. Ex.: Hydnum.

Family 2. Polyporaceæ, or pore fungi. Ex.: Polyporus.

Family 3. Agaricaceæ, the gill family in which the hymenium covers blade-like plates of the pileus, called gills, generally occurring on the under surface of the same. Ex.: Agaricus campestris, the common edible mushroom of fields; Anamita muscaria and Anamita phalloides, both of which are poisonous.

Division b.—Gasteromycetes

(Hymenium inclosed)

Order 1. Lycoperdales, or puff balls. Ex.: Geaster, the earth star and Lycoperdon.

Order 2. Nidulariales, the nest fungi.

SUBDIVISION V.—LICHENES, THE LICHENS

Lichens are variously colored, usually dry and leathery plants, consisting of symbioses of algæ and fungi. They are found on the bark of trees, on rocks, logs, old fences, etc.

According to structure and mode of growth of the thallus, the

Lichenes are, like the Fungi, divided into several sub-groups. A perfect lichen usually consists of a thallus, or vegetable apparatus; apothecia, or organs of fructification, and spermogonia, or organs of fertilization.

DIVISION II.—BRYOPHYTA

Plants showing a beginning of definite alternation of generations, *i.e.*, gametophyte (sexual phase) alternating with sporophyte (asexual phase of development) in their life history, the two phases being combined in one plant.

SUBDIVISION I.—HEPATICÆ OR LIVERWORTS

Plants of aquatic or terrestrial habit whose bodies consist of a rather flat, furchate branching thallus or leafy branch which is dorsiventral (having distinct upper and lower surface); the upper surface consists of several layers of cells containing chlorophyll, which gives the green color to the plants; the lower surface gives origin to hair-like outgrowths of the epidermal cells serving as absorptive parts and called rhizoids. Upon the dorsal surface of this thalloid body (the gametophyte) cup-like structures are produced called cupules which contain special reproductive bodies called gemmæ, these being able to develop into new gametophytes. The sex organs are of two kinds, male and female. The male organs are termed antheridia, the female, archegonia. The antheridia are more or less club-shaped, somewhat stalked organs consisting of an outer layer of sterile cells investing a mass of sperm mother cells from which are formed the spirally curved biciliate antherozoids, or male sexual cells. The archegonia are flask-shaped organs consisting of an investing layer of sterile cells surrounding an axial row of cells, the neck canal cells, ventral canal cells and the egg or female sexual cell. Every cell of the axial row breaks down in the process of maturation with the exception of the egg which remains in the basal portion. Both antheridia and archegonia generally arise on special stalks above the dorsal surface. After the egg is fertilized by an antherozoid, the young embryo resulting grows into a sporogonium (the sporophyte) consisting of a stalk portion partly imbedded in the archegonium surmounting a sporangium or capsule in which spores are produced. When mature the capsule splits open discharging the spores. The spores on germination develop into a protonema or filamentous outgrowth which later develops the thallus.

Order 1. **Marchantiales,** including Marchantia and Riccia.

Order 2. **Jungermaniales,** the leafy liverworts, including Porella.

Order 3. **Anthocerotales,** having the most complex sporophyte generations among liverworts, including Anthoceros, and Megaceros.

SUBDIVISION II.—MUSCI OR MOSSES

Plants found on the ground, on rocks, trees and in running water. Their life histories consist of two generations, gametophyte and sporophyte similar to the liverworts but differ from liverworts, generally, by the ever-present differentiation of the gametophyte body into distinct stem and simple leaves, and the formation of the sexual organs at the end of an axis of a shoot. They are either monœcious, when both kinds of sexual organs are borne on the same plant, or diœcious, in which case the antheridia and archegonia arise on different plants.

Order 1. **Sphagnales,** or Bog Mosses, including the simple genus, Sphagnum. Pale mosses of swampy habit whose upper extremities repeat their growth periodically while their lower portions die away gradually and form peat, hence their frequent name of Peat Mosses.

Order 2. **Andreæales,** including the single genus Andreæa, a xerophytic habit occurring on siliceous rock.

Order 3. **Bryales,** or true mosses comprising the most highly evolved type of bryophytes. Ex.: Polytrichum, Funaria, Hypnum, and Minium.

Life History of Polytrichum Commune (a Typical True Moss)

Polytrichum commune is quite common in woods, forming a carpet-like covering on the ground beneath tall tree canopies. It is diœcious, the plants being of two kinds, male and female.

Beginning with a spore which has fallen to the damp soil, we note its beginning of growth (germination) as a green filamentous body called a protonema. This protonema soon becomes branched, giving rise to hair-like outgrowths from its lower portion called rhizoids and lateral buds above these which grow into leafy stems commonly known as "moss plants." At the tips of some of these leafy stems antheridia (male sexual organs) are formed while on others archegonia (female sexual organs) are formed. These organs are surrounded at the tips by delicate hairy processes called paraphyses as well as leaves for protection. The antheridia bear the antherozoids, the archegonia, the eggs or ova, as in the liverworts. When an abundance of moisture is present the antherozoids are liberated from the antheridia, swim through the water to an archegonium and descend the neck canal, one fertilizing the egg by uniting with it. This completes the sexual or gametophyte generation. The fertilized egg now undergoes division until an elongated stalk bearing upon its summit a capsule is finally produced, this being known as the sporogonium. The base of the stalk remains imbedded in the basal portion of the arche-gonium at the tip of the leafy stalk and forms a foot or absorbing process. In growing upward the sporogonium ruptures the neck of the archegonium and carries it upward as the covering of the capsule, or calyptra. The calyptra is thrown off before the spores are matured within the capsule. The upper part of the capsule becomes converted into a lid or operculum at the margin of which an annulus or ring of cells forms. The cells of the annulus are hygroscopic and expand at maturity, throwing off the lid and allowing the spores to escape. This completes the asexual or sporophyte generation. The spores falling to the damp soil germinate into protonemata, thus completing the life cycle in which is seen an alteration of genera-tions, the two phases, gametophyte alternating with sporophyte.

DIVISION III.—PTERIDOPHYTA

The most highly developed cryptogams showing a distinct alter-nation of generations in their life history. They differ from the Bryophytes in presenting independent, leafy, vascular, root-bearing sporophytes.

SUBDIVISION I.—LYCOPODIALES OR CLUB MOSSES

Small perennial vascular, dichotomously branched herbs with stems thickly covered with awl-shaped leaves. The earliest forms of vascular plants differing from ferns in being comparatively simple in structure, of small size, leaves sessile and usually possessing a single vein. Except in a few instances the sporangia are borne on leaves, crowded together and forming cones or spikes at the ends of the branches. Homosporous.

FAMILY 1. LYCOPODIACEÆ, including the single genus Lycopodium with widely distributed species. The spores of Lycopodium clavatum are official.

FAMILY 2. SELAGINELLACEÆ, including the single genus Selaginella, with species for the greater part tropical. Plants similar in habit to the Lycopodiaceæ but showing heterospory.

FAMILY 3. ISOETACEÆ, including the single genus Isoetes whose species are plants with short and tuberous stems giving rise to a tuft of branching roots below and a thick rosette of long, stiff awl-shaped leaves above. Heterosporous.

SUBDIVISION II.—EQUISETALES

(The Horsetails or Scouring Rushes)

The Equisetales, commonly known as the Horsetails or Scouring rushes are perennial plants with hollow, cylindrical, jointed and fluted stems, sheath-like whorls of united leaves and terminal cone-like fructifications. Their bodies contain large amounts of silicon, hence the name scouring rushes.

In some varieties the fruiting cone is borne on the ordinary stem, in others on a special stem of slightly different form. In the latter the spores are provided with elaters, which, being hygroscopic, coil and uncoil with increase or decrease in the amount of moisture present, thus aiding in the ejection of spores from the sporangia. The number of species is small and included under one genus Equisetum.

SUBDIVISION III.—FILICALES

The group Filicales is the largest among the vascular cryptogams and includes all the plants commonly known as Ferns. The main axis of a typical fern is a creeping underground stem or rhizome which at its various nodes bears rootlets below and fronds above. These fronds are highly developed, each being provided with a petiole-like portion called a stipe which is extended into a lamina usually showing a forked venation. Some ferns possess laminæ which are lobed, each lobe being called a pinna. If a pinna be further divided, its divisions are called pinnules. The unfolding of a frond is circinate and it increases in length by apical growth. On the under surface of the laminæ, pinnæ, or pinnules may be seen small brown patches each of which is called a sorus, and usually covered by a membrane called the indusium. Each sorus consists of a number of sporangia (spore cases) developed from epidermal cells. In some ferns the entire leaf becomes a spore-bearing organ (sporophyll). Most sporangia have a row of cells around the margin, the whole being called the annulus. Each cell of the annulus has a U-shaped thickened cell wall. Water is present within these cells and when it evaporates it pulls the cell walls together, straightening the ring and tearing open the weak side. The annulus then recoils and hurls the spores out of the sporangium. Upon coming in contact with damp earth each spore germinates, producing a green septate filament called a protonema. This later becomes a green heart-shaped body called a prothallus. It develops on its under surface antheridia or male organs and archegonia or female organs as well as numerous rhizoids. Within the antheridia are developed motile sperm, while ova are produced within the archegonia. The many ciliate sperms escape from the antheridia of one prothallus during a wet season and moving through the water are drawn by a chemotactic influence to the archegonia of another prothallus, pass down the neck canals of these and fuse with the ova, fertilizing them. The fertilized egg or

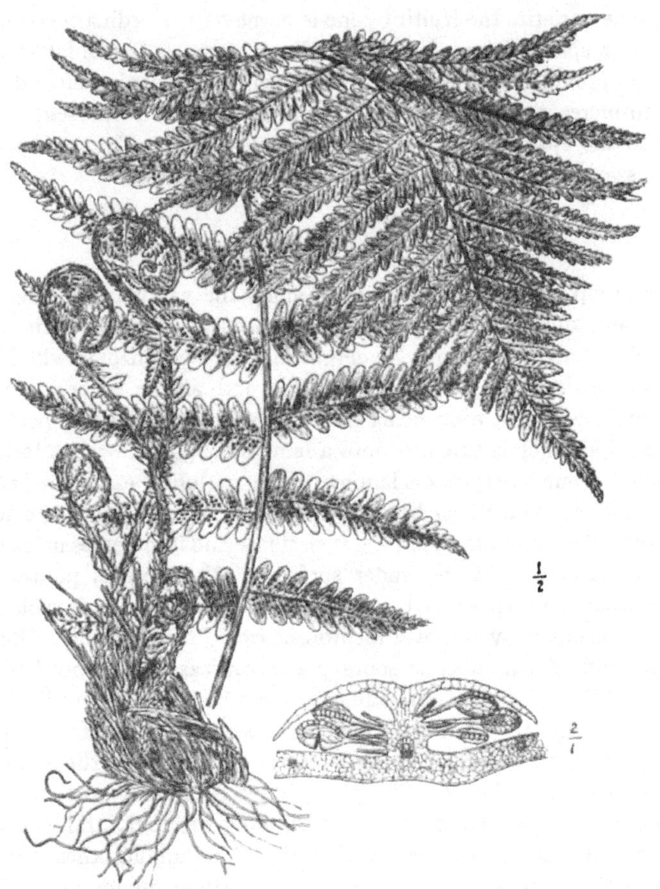

Fig. 37.—*Dryopteris filix-mas*—Plant and section through sorus. (*From Sayre.*)

oöspore divides and redivides and soon becomes differentiated into stem bud, first leaf, root, and foot. The foot obtains nourishment from the prothallus until the root grows into the soil, when it atrophies and the sporophyte becomes independent. Unequal growth and division of labor continue until a highly differentiated sporophyte results, the mature "fern plant."

CLASS 1. FILICINEÆ OR TRUE FERNS (HOMOSPOROUS)

FAMILY POLYPODIACEÆ.—Sporangia with annulus vertical and incomplete.

The rhizomes of Dryopteris filix-mas and Dryopteris marginalis are official in the U. S. P. The fibrovascular bundles of these are concentric in type but differ from the concentric f.v. bundles of some monocotyledons in that xylem is innermost and phloem surrounds the xylem.

CLASS 2. HYDROPTERIDINEÆ, OR WATER FERNS (HETEROSPOROUS)

FAMILY SALVINIACEÆ, floating ferns with broad floating leaves and submerged dissected leaves which bear sporocarps. Ex.: Salvinia and Azolla.

DIVISION IV.—SPERMATOPHYTA (PHANEROGAMIA)

Plants producing real flowers and seeds. The highest evolved division of the vegetable kingdom.

SUBDIVISION I.—ANGIOSPERMIA OR ANGIOSPERMS

(Plants with covered seeds)

CLASS A.—MONOCOTYLEDONS

A class of Angiospermia characterized by the following peculiarities:

One cotyledon or seed leaf in the embryo.

Stems endogenous with closed collateral or concentric fibrovascular bundles, which are scattered.

Leaves generally parallel veined.

Flowers trimerous (having the parts of each whorl in 3's or multiple thereof).

Secondary growth in roots generally absent.

Medullary rays generally absent.

FAMILY 1. ARACEÆ OR ARUM FAMILY.—Perennial herbs with fleshy rhizomes or corms, and long petioled leaves, containing an acrid or pungent juice. Flowers crowded on a spadix, which is usually surrounded by a spathe. Fruit a berry. Seeds with large fleshy embryo.

Official drug	Part used	Botanical name
Calamus	Unpeeled rhizome	Acorus calamus
Unofficial drug		
Skunk cabbage	Rhizome	Symplocarpus fœtidus
Indian turnip	Corm	Arisæma triphyllum

FAMILY 2. GRAMINACEÆ OR GRASS FAMILY.—Mostly herbs with cylindric, hollow jointed stems whose nodes are swollen. The leaves are alternate, with long split sheaths and a ligule. Flowers generally hermaphroditic and borne in spikelets making up a spicate inflorescence. Lowest floral leaves of each spikelet are called glumes, which are empty and paired. Fruit, a caryopsis or grain. Embryo with scutellum.

Official drug	Part used	Botanical name
Triticum	Rhizome	Agropyron repens
Saccharum	Refined sugar	Saccharum officinarum and Sorghum sp.?
Maltum	Seed, partially germinated and dried	Hordeum distichum
Zea	Styles and stigmas	Zea mays

FAMILY 3. PALMEÆ OR PALM FAMILY.—Tropical or sub-tropical arborescent plants, having unbranched trunks which are terminated by a crown of leaves, in the axils of which the flowers are produced. The leaves are well developed with pinnate or palmate blades and a fibrous sheathed clasping petiole. Inflorescence lateral with small flowers. Fruit a berry or drupe.

Official drug	Part used	Botanical name
Sabal	Fruit	Serenoa serrulata
Unofficial		
Cocoanut oil	Fixed oil	Cocos nucifera
Carnauba wax	Wax from leaves	Copernica cerifera
Areca nut	Seed	Areca Catechu

FAMILY 4. LILIACEÆ OR LILY FAMILY.—Herbs with regular and symmetrical almost always six-androus flowers. The perianth is parted into six segments, the calyx and corolla being alike in color. Anthers introrse. Ovary three-locular with a single style. Fruit a capsule or berry.

Official drug	Part used	Botanical name
Sarsaparilla	Root	Smilax medica Smilax ornata Smilax papyraceæ Smilax officinalis
Convallaria	Rhizome and roots	Convallaria majalis
Veratrum	Rhizome and roots	Veratrum viride Veratrum album
Colchici Cormus	Corm	Colchicum autumnale
Colchici Semen	Seed	Colchicum autumnale
Aloe	Inspissated juice of leaves	Aloe vera Aloe Perryi Aloe chinensis Aloe ferox
Scilla	Bulb	Urginea maritima
Veratrina	Mixture of alkaloids	Asagræa officinalis
Unofficial		
Garlic	Bulb	Allium sativum

FAMILY 5. IRIDACEÆ OR IRIS FAMILY.—Perennial herbs with equitant two-ranked leaves and regular or irregular flowers which are showy. Fruit a three-celled, loculicidal, many-seeded capsule. Rootstocks, tubers, or corms mostly acrid.

Unofficial drug	Part used	Botanical name
Orris	Rhizome	Iris florentina Iris pallida Iris germanica
Saffron	Stigmas	Crocus sations

FAMILY 6. ORCHIDACEÆ OR ORCHID FAMILY.—Perennial herbs having grotesque flowers. The perianth consists of six segments, the outer three of which correspond to sepals and are similar. Two segments of the inner circle resemble petals while the third is known as the Labellum or Lip. This is generally larger than the rest and is directed downward and forward. Usually only one stamen is developed, bearing anther. Leaves parallel veined and alternate. Ovary unilocular with many ovules on three parietal placentæ. Fruit a capsule, three-valved and one-celled. Seeds minute.

Official drug	Part used	Botanical name
Vanilla	Fruit	Vanilla planifolia
Cypripedium	Rhizome and roots	Cypripedium hirsutum Cypripedium parviflorum

Family 7. Zingiberaceæ or Ginger Family.—Tropical plants, perennial herbs with fleshy rhizomes and large elliptical pinnately veined leaves. The leaf sheaths are folded tightly around each other so as to give the appearance of a stem. Flowers, zygomorphic.

Official drug	Part used	Botanical name
Zingiber	Rhizome	Zingiber officinale
Cardamomum	Fruit	Elettaria repens

Class B.—Dicotyledons

Plants having the following characteristics:

Two-seed leaves (cotyledons) in embryo.

Netted veined leaves.

Open collateral fibrovascular bundles, radially arranged about pith.

Exogenous stems.

Medullary rays present.

Cambium.

Roots developing secondary structure.

Flowers tetra- or pentamerous (parts of each whorl, four or five or multiple thereof.

Sub-class a.—Archichlamydeæ

Those dicotyledonous plants in which the petals are distinct and separate from one another or are entirely wanting. That group of the Archichlamydeæ whose flowers show the absence of petals and frequently of sepals is called the Apetalæ. The group whose plants have flowers showing the parts of their corolla (petals) separate and distinct is called the Chloripetalæ.

The Apetalæ

Family 1. Piperaceæ or Pepper Family.—A family of aromatic herbs and shrubs with jointed stems, opposite, verticillate, or sometimes alternate leaves without stipules, and spiked or racemose flowers.

Official drug	Part used	Botanical name
Cubeba	Unripe fruit	Piper Cubeba
Piper	Unripe fruit	Piper nigrum
Matico	Leaves	Piper angustifolium
Unofficial		
Methysticum	Root	Piper methysticum

FAMILY 2. FAGACEÆ OR BEECH FAMILY (Cupuliferæ).—Apetalous trees or shrubs having alternate pinnately veined leaves, monœcious flowers, the male in drooping aments, the female solitary, clustered, or in scaly catkins. Fruit a one-celled one-seeded nut. The beech, oak and chestnut, are the principal genera.

Official drug	Part used	Botanical name
Galla	Excrescence	Quercus infectoria
Creosotum	Product of distillation	{ Fagus ferruginea
Quercus	Bark	{ Fagus sylvatica, etc.
Unofficial		Quercus alba
Castanea	Leaves	Castanea dentata

The cork of commerce is obtained from the bark of Quercus Suber and Quercus occidentalis, plants indigenous to Spain and France.

FAMILY 3. BETULACEÆ OR BIRCH FAMILY.—A family of trees or shrubs distinguished by monœcious flowers with scaly bracts and astringent resinous bark. Differs from Fagaceæ by superior ovary and absence of cupule. To this family belong the hazelnuts, birches, alders, the ironwood, and the hornbean.

Official drug	Part used	Botanical name
Oleum Betulæ	Volatile oil	Betula lenta

FAMILY 4. JUGLANDACEÆ.—A family of apetalous exogenous trees —the walnut family—with alternate odd-pinnate leaves and monœcious flowers, the sterile in catkins, the fertile solitary or in a small cluster or spike. The fruit is a dry drupe with a bony nut-shell and a four-lobed seed. It embraces five genera, of which *Cary* (Hicoria) and *Juglans* are represented in the United States, and about 35 species.

Unofficial drug	Part used	Botanical name
Juglans	Root bark	Juglans cinerea

FAMILY 5. SALICACEÆ OR WILLOW FAMILY.—A family of apetalous shrubs or trees—the willow family—having alternate undivided leaves and diœcious flowers (one to each bract) in catkins. It embraces two genera—*Salix*, the willows, and *Populus*, the poplars—and from 180 to 300 species, found chiefly in northern temperate and frigid regions, there being none in Australia or the South Pacific islands.

Official drug	Part used	Botanical name
Salicin	Glucoside	Several species of Salix and Populus

FAMILY 6. MYRISTICACEÆ.—An order of apetalous trees—the nutmeg family—comprising the single genus *Myristica*, of about 80 species.

Myristica.—A large tropical genus of fragrant, apetalous trees— the nutmegs—coextensive with the nutmeg family, having alternate, entire, often punctate leaves, small diœcious regular flowers, and a succulent, two-valved one-celled fruit with a solitary seed usually covered by a lancinate aril.

M. fragrans, a handsome tree, 20 to 30 feet high, of the Malay archipelago, supplies the nutmegs and mace of commerce.

Official drug	Part used	Botanical name
Myristica	Kernel of seed	Myristica fragrans
Oleum Myristicæ	Volatile oil	Myristica fragrans

FAMILY 7. LAURACEÆ OR LAUREL FAMILY.—A family of aromatic trees or shrubs with alternate, coriaceous, pellucid punctate leaves containing considerable volatile oil; flowers polygamous, each having a calyx of four or six colored sepals.

Official drug	Part used	Botanical name
Camphora	Stearopten	Cinnamomum Camphora
Sassafras	Bark of root	Sassafras variifolium
Sassafras Medulla	Pith	Sassafras variifolium
Cinnamomum Zeylanicum	Bark	Cinnamomum zeylanicum
Cinnamomum Saigonicum	Bark	Undetermined species of cinnamon
Oleum Cinnamomi	Volatile oil	Cinnamomum cassia
Unofficial		
Coto	Bark	Drimys winteri
Laurus	Leaves	Laurus nobilis

FAMILY 8. MYRICACEÆ, OR BAYBERRY FAMILY.—A family of evergreen or deciduous, apetalous, mostly diœcious shrubs and trees included within the single genus, Myrica. Flowers in mostly single, seldom closely set aments, leaves single, occasionally (Myrica asplenifolia) pinnately cleft. Fruit, a waxy drupe.

The outer waxy layer of the fruit is used in making a superior candle while an infusion or fluid extract of the bark and leaves is used as a specific in various affections of the mucous membranes, diarrhœa, dysentery, etc.

FAMILY 9. POLYGONACEÆ OR BUCKWHEAT FAMILY.—Apetalous herbs, shrubs, or rarely trees with alternate entire leaves, the stipules forming a sheath above the swollen joints of the stem; flowers, small and with a two- to six-parted perianth; fruit, an angled akene.

Official drug	Part used	Botanical name
Rheum	Rhizome	Rheum officinale Rheum palmatum and the variety tanguticum
Unofficial		
Rumex	Root	Rumex crispus

FAMILY 10. PHYTOLACCACEÆ.—A family of apetalous trees, shrubs, or woody herbs—the pokeweed family—with alternate entire leaves and flowers resembling those of the goosefoot family (*Chenopodiaceæ*), but differing in having the several-celled ovary composed of carpels united in a ring, and forming a berry in fruit. It embraces 21 genera, and 55 species, tropical and sub-tropical.

Official drug	Part used	Botanical name
Phytolacca	Root	Phytolacca decandra

FAMILY 11. CHENOPODIACEÆ.—A family of more or less succulent apetalous annual or perennial herbs—the goosefoot family—with usually alternate exstipulate leaves and minute greenish flowers. It embraces about 80 genera and over 500 species, among them being several garden vegetables and a number of weeds.

Official drug	Part used	Botanical name
Oleum Chenopodii	Volatile oil	Chenopodium anthelminticum
Saccharum	Refined sugar	Beta vulgaris

FAMILY 12. ARISTOLOCHIACEÆ.—A small family of apetalous plants—the birthwort family—chiefly climbers or twiners and tropical, with irregular, dingy, often offensively smelling flowers. There are five genera and about 200 species.

Official drug	Part used	Botanical name
Serpentaria	Rhizome and roots	Aristolochia serpentaria Aristolochia reticulata
Unofficial		
Asarum	Rhizome and roots	Asarum canadensis

FAMILY 13. ULMACEÆ OR ELM FAMILY.—Forest trees indigenous to the temperate and tropical zones, characterized by being woody plants, with pinnately veined leaves and caducous stipules and without

milky juice. Their flowers are unisexual or hermaphroditic with six or four parts to the perianth. Fruit, a samara.

Official drug	Part used	Botanical name
Ulmus	Inner bark	Ulmus fulva

FAMILY 14. MORACEÆ OR MULBERRY FAMILY.—Mostly shrubs or trees, rarely herbs, perennials, with small axillary, clustered or solitary unisexual flowers, variously colored; leaves ovate with serrate margin and having caducous stipules; fruit an akene enclosed by the perianth. Milky juice present.

Official drug	Part used	Botanical name
Cannabis Indica	Flowering tops of pistil- late plant	Cannabis sativa
Ficus	Fruit	Ficus carica
Humulus	Strobile	Humulus lupulus
Lupulinum	Glandular trichome	Humulus lupulus

FAMILY 15. EUPHORBIACEÆ OR SPURGE FAMILY.—A vast group of apetalous trees, shrubs, or herbs mainly natives of warm countries, with milky acrid juice, normally alternate, entire leaves; fruit, a three-locular capsule containing seeds with oily endosperm. Some plants furnish rubber.

Official drug	Part used	Botanical name
Elastica	Concrete juice	Hevea species
Stillingia	Root	Stillingia sylvatica
Oleum Ricini	Volatile oil	Ricinus communus
Oleum Tiglii	Volatile oil	Croton tiglium
Unofficial		
Cascarilla	Bark	Croton eluteria
Tapioca	Starch	Manihot utilissima
Kamala	Hairs of capsule	Mallotus philipinensis

The Chloripetalæ (Polypetalæ)

Flowers have both calyx and corolla, the latter being composed of distinct petals.

FAMILY 16. MAGNOLIACEÆ OR MAGNOLIA FAMILY.—Trees and shrubs having alternate leaves and single large flowers with calyx and corolla colored alike. Sepals and petals deciduous, anthers adnate. Pistils and stamens numerous. Bark aromatic and bitter.

Official drug	Part used	Botanical name
Oleum anisi	Volatile oil	Illicium verum

Family 17. Rosaceæ.—A family of polypetalous plants—the rose family—with alternate simple or compound stipulate leaves, and regular flowers with usually numerous distinct stamens inserted on the urn-shaped calyx. It embraces 80 genera, and nearly 2000 species, found in all parts of the world.

Trees, shrubs and a few herbs. The flowers bear comparatively many petals. The fruits vary greatly and may be fleshy, an akene, berry or a drupe. Many of the fruits are edible.

Official drug	Part used	Botanical name
Oleum Rosæ	Volatile oil	Rosa damascena
Amygdala Amara	Seed	Prunus amygdalus variety amara
Amygdala Dulcis	Seed	Prunus amygdalus variety dulcis
Prunus Virginiana	Bark	Prunus serotina
Rubus	Bark	Rubus villosus, R. cuneifolius, and R. nigrobaccus
Quillaja	Bark	Quillaja saponaria
Cusso	Panicles of pistillate flowers	Hagenia abyssinica
Rosa Gallica	Petals	Rosa gallica
Unofficial		
Laurocerasus	Leaves	Prunus laurocerasus
Cydonium	Seed	Cydonia vulgaris

Family 18. Anacardiaceæ, or Cashew Family.—A family of chloripetalous trees or shrubs, with resinous, acrid, milky juice, alternate leaves, small flowers, and a mostly drupaceous fruit. Exhalations of many members frequently poisonous especially from the Rhus venenata, and R. Toxicodendron (Poison Ivy).

Official drug	Part used	Botanical name
Mastiche	Resinous exudation	Pistacia lentiscus
Rhus glabra	Fruit	Rhus glabra
Unofficial		
Rhus Toxicodendron	Leaves	Rhus toxicodendron
Rhus Aromatica	Bark of root	Rhus aromatica

Family 19. Ranunculaceæ.—An order of herbaceous or woody plants—the crowfoot or buttercup family—with radical or alternate palmately veined leaves, and terminal, racemose, or panicled flower clusters, the flowers regular or irregular, with all parts distinct and unconnected. There are 30 genera and 1350 species.

Medium-sized shrubs or herbs with acrid juices. Fruit is an akene, pod or berry. Chiefly temperate or cold climates. Seeds contain albuminous matter.

Official drug	Part used	Botanical name
Hydrastis	Rhizome and roots	Hydrastic canadensis
Aconitum	Tuberous root	Aconitum napellus
Staphisagria	Seed	Delphinium Staphisagria
Cimicifuga	Rhizome and root	Cimicifuga racemosa
Unofficial		
Pulsatilla	Entire herb	{ Anemone pulsatilla { Anemone pratensis
Coptis	Entire herb	Coptis trifolia
Helleborus	Rhizome and roots	Helleborus niger
Adonis	Entire herb	Adonis vernalis

FAMILY 20. LEGUMINOSÆ.—A vast family of polypetalous herbs, shrubs and trees—the bean family—with alternate, stipulate, usually compound leaves and papilionaceous or sometimes regular flowers, with usually 10 monadelphous, diadelphous, or rarely distinct stamens, and a simple pistil becoming generally a legume in fruit. It embraces three well-marked groups, 24 tribes, 427 genera, and 7000 species.

Official drug	Part used	Botanical name
Acacia	Gummy exudation	Acacia Senegal and other species
Tragacantha	Gummy exudation	Astragalus gummifer and other species
Balsamum Peruvianum	Balsam	Toluifera pereiræ
Balsamum Tolutanum	Balsam	Toluifera Balsamum
Hæmatoxylon	Heartwood	Hæmatoxylon campechianum
Santalum Rubrum	Heartwood	Pterocarpus santalinus
Glycyrrhiza	Rhizome and root	{ Glycyrrhiza glabra { Glycyrrhiza glandulifera
Senna	Leaflets	{ Cassia acutifolia { Cassia angustifolia
Cassia Fistula	Fruit	Cassia fistula
Tamarindus	Pulp of fruit	Tamarindus indica
Copaiba	Oleoresin	Copaiba species
Chrysarobinum	Neutral principle	Vouacapoua araroba
Physostigma	Seed	Physostigma venenosum
Kino	Inspissated juice	Pterocarpus Marsupium
Scoparius	Tops	Cytisus Scoparius
Unofficial		
Fœnum græcum	Seed	Trigonella fœnum-græcum
Piscidia	Bark	Piscidia erythrina
Indigo	Coloring matter	Indigofera tinctoria
Trifolium	Flower heads	Trifolium pratense
Dipteryx	Fruit	Dipteryx odorata

FAMILY 21. CRUCIFERÆ OR MUSTARD FAMILY.—A large family of annual or perennial polypetalous herbs with pungent watery juice and cruciform corollas; stamens tetradynamous; fruit a silique.

Official drug	Part used	Botanical name
Sinapis Alba	Seed	Sinapis alba
Sinapis Nigra	Seed	Brassica nigra

FAMILY 22. BURSERACEÆ.—A small family of tropical balsamiferous or resinous polypetalous trees or shrubs—the myrrh family—with alternate compound leaves and three to five parted usually perfect flowers. It includes 18 genera and 150 species.

Official drug	Part used	Botanical name
Myrrha	Gum resin	Commiphora Myrrh
Unofficial		
Olibanum	Gum resin	Boswellia carterii

FAMILY 23. CACTACEÆ.—A family of American polypetalous plants—the cactus family—green and fleshy, and mostly leafless, having globular or columnar, tuberculated or ribbed, or jointed and often flattened stems, usually armed with bundles of spines, and bearing large and often showy flowers with numerous sepals, petals and stamens, and the fruit a pulpy berry. It embraces 13 genera and about 1000 species.

Unofficial drug	Part used	Botanical name
Cactus	Fresh branches	Cereus grandiflorus

FAMILY 24. BERBERIDACEÆ OR BARBERRY FAMILY.—Herbs and woody plants with watery juices and alternate, or radical, simple or compound leaves often bearing spines or barbs, which give them a barbed appearance. Fruit a berry or capsule.

Official drug	Part used	Botanical name
Berberis	Rhizome and roots	Berberis aquifolium and other species
Podophyllum	Rhizome	Podophyllum peltatum

FAMILY 25. CELASTRACEÆ.—A family of polypetalous trees or shrubs—the staff-tree or spindle-tree family—mostly tropical, having simple, coriaceous leaves, small regular flowers with imbricated sepals and petals, and four or five perigynous stamens inserted on a fleshy disc alternately with the petals; seed in a succulent aril.

Official drug	Part used	Botanical name
Euonymus	Bark	Euonymus atropurpureus

FAMILY 26. CANELLACEÆ.—A small family of tropical American polypetalous, aromatic trees—the canella family—with alternate, exstipulate, entire leaves and axillary, cymose, perfect flowers. It embraces two genera, *Canella* and *Cinnamodendron*, and about four species.

Unofficial drug	Part used	Botanical name
Canellæ cortex	Bark	Canella alba

FAMILY 27. CORNACEÆ.—A family of polypetalous shrubs or trees —the dogwood or comel family—of all parts of the world, with usually alternate coriaceous entire leaves, and terminal or axillary cymose clusters of small flowers. It embraces 15 genera and 80 species.

Unofficial drug	Part used	Botanical name
Cornus	Bark	Cornus florida

FAMILY 28. THYMELEACEÆ.—A family of trees or shrubs, the spurge laurel or mezereum family, having very tough bark, opposite entire leaves and small, perfect, regular flowers.

Official drug	Part used	Botanical name
Mezereum	Bark	Daphne Mezereum

FAMILY 29. GERANIACEÆ.—A family of polypetalous herbs, shrubs or trees—the geranium family—usually with lobed or dissected leaves and axillary peduncles of often showy, perfect flowers. It embraces seven tribes, 25 genera, and about 980 species, widely scattered in temperate and sub-tropical regions.

Official drug	Part used	Botanical name
Geranium	Rhizome	Geranium maculatum

FAMILY 30. HAMAMELIDACEÆ.—A family of polypetalous shrubs or trees—the witch-hazel family—with alternate simple leaves, two deciduous stipules, and heads or spikes of monœcious or polygamous flowers. It includes 19 genera and about 40 species.

The leaves and twigs contain highly aromatic volatile oils.

Official drug	Part used	Botanical name
Hamamelidis folia	Leaves	Hamamelis virginiana
Hamamelidis cortex	Bark and twigs	Hamamelis virginiana
Styrax	Balsam	Liquidambar orientalis

FAMILY 31. LINACEÆ.—A family of polypetalous herbs, shrubs, or rarely trees—the flax family—with alternate simple and usually entire

leaves, and regular, symmetrical, hypogynous flowers which are four-
to five-membered throughout, the petals blue, yellow, or white, and
fugacious. It embraces 15 genera and about 235 species, distributed
over the world.

Official drug	Part used	Botanical name
Linum	Seed	Linum usitatissimum

FAMILY 32. GUTTIFERÆ.—A family of polypetalous trees or shrubs
—the gamboge family—with resinous juice, opposite, coriaceous leaves,
and terminal or axillary clusters of regular diœcious flowers. It embraces 26 genera and about 370 species, all natives of the tropics.

Official drug	Part used	Botanical name
Cambogia	Gum resin	Garcinia hanburii

FAMILY 33. MALVACEÆ OR MALLOW FAMILY.—A family of chloripetalous herbs, shrubs, or trees abounding in mucilage and usually
with the above-ground portion covered with trichomes; the leaves are
alternate and palmately nerved; the flowers regular, the corolla beau-
tifully colored, funnel or bell-shaped, stamens monadelphous; fruit a
several-celled pod.

Official drug	Part used	Botanical name
Althæa	Root (peeled)	Althæa officinalis
Gossypii Cortex	Bark of root	Gossypium herbaceum
Gossypium Purificatum	Hairs of seed	Gossypium herbaceum
Oleum Gossypii seminis	Oil of seed	Gossypium herbaceum

FAMILY 34. PAPAVERACEÆ.—A family of polypetalous plants—
the poppy family—usually with milky or colored juice, alternate exstipulate leaves, and long one-flowered peduncles, the flowers usually
with two caducous sepals and four cruciate petals. It embraces
about 20 genera and 80 species.

Herbs or low shrubs with milky or colored, narcotic juices. Flowers
showy. Fruit usually a many-sided capsule. Temperate and tropical
regions.

Official drug	Part used	Botanical name
Opium	Concrete milky exudate	Papaver somniferum
Sanguinaria	Rhizome	Sanguinaria canadensis

FAMILY 35. PASSIFLORACEÆ.—A family of polypetalous shrubs,
trees, or rarely herbs—the passion-flower family—often climbing, with
alternate, palmately lobed or compound leaves and solitary or racemose,

often handsome, flowers with five monadelphous stamens. It embraces five tribes, 27 genera, and 235 species, all tropical or sub-tropical.

Unofficial drug	Part used	Botanical name
Papain	Ferment	Carica papaya
Passiflora	Rhizome	Passiflora incarnata

FAMILY 36. MENISPERMACEÆ, OR MOONSEED FAMILY.—Chloripetalous woody, climbing tropical plants with alternate simple leaves; flowers green to white; fruit a one-seeded succulent drupe. They usually contain tonic, narcotic or poisonous bitter principles.

Official drug	Part used	Botanical name
Calumba	Root	Jateorhiza palmata
Pareira	Root	Chondodendron tomentosum
Unofficial		
Cocculus	Fruit	Anamirta paniculata
Menispermum	Rhizome and roots	Menispermum canadense

FAMILY 37. MYRTACEÆ OR MYRTLE FAMILY.—Evergreen trees or shrubs of warmer climates, with opposite, entire exstipulate leaves of an elliptical shape and having a vein running close to the margin. All the organs provided with roundish glands containing hydrocarbon principles, giving them an aromatic odor. Flowers with imbricate calyx lobes, numerous stamens and an inferior ovary.

Official drug	Part used	Botanical name
Eucalyptus	Leaves	Eucalyptus globulus
Eucalyptol	Organic oxide	Eucalyptus globulus
Caryophyllus	Flower bud	Eugenia aromatica
Eugenol	Aromatic phenol	Eugenia aromatica
Pimenta	Fruit	Pimenta officinalis
Unofficial		
Myrcia	Volatile oil and leaves	Myrcia acris

FAMILY 38. POLYGALACEÆ.—A family of polypetalous herbs, shrubs, or rarely small trees—the milkwort family—having alternate simple entire leaves and irregular hypogynous flowers with four to eight diadelphous or monadelphous stamens.

Official drug	Part used	Botanical name
Senega	Root	Polygala Senega

FAMILY 39. RUTACEÆ OR RUE FAMILY.—A family of pellucid, punctate, polypetalous woody plants, rarely herbs having exstipulate opposite, simple or compound leaves and variously shaped inflorescences

of perfect, five-parted flowers; fruit a capsule or berry. The plants contain ethereal oils in their intracellular cavities.

Official drug	Part used	Botanical name
Aurantii Dulcis Cortex	Outer rind of ripe fruit	Citrus Aurantium
Aurantii Amari Cortex	Rind of unripe fruit	Citrus vulgaris
Limonis Cortex	Outer rind of ripe fruit	Citrus Limonum
Limonis Succus	Fresh juice of ripe fruit	Citrus Limonum
Pilocarpus	Leaflets	{ Pilocarpus Jaborandi { Pilocarpus microphyllus
Buchu (short)	Leaves	Barosma betulina
Xanthoxylum	Bark	{ Xanthoxylum americanum { Fagara Clava-Herculis

FAMILY 40. RHAMNACEÆ OR BUCKTHORN FAMILY.—Chloripetalous shrubs or small trees of warm temperate regions with spiny stems, simple leaves, small regular flowers, and fleshy winged drupaceous fruit.

Official drug	Part used	Botanical name
Rhamnus Purshiana	Bark	Rhamnus Purshiana
Frangula	Bark	Rhamnus Frangula
Unofficial		
Rhamnus Cathartica	Fruit	Rhamnus cathartica

FAMILY 41. TURNERACEÆ.—An order of polypetalous herbs or shrubs—the turnerad family—mainly American, having alternate simple or pinnalified leaves, and axillary solitary or few-clustered perfect flowers with five stamens. There are six genera and 85 species. *Turnera*, the type genus, furnishes a number of ornamental greenhouse plants and the drug damiana.

Unofficial	Part used	Botanical name
Damiana	Leaves	Turnera diffusa variety aphrodisiaca

FAMILY 42. SIMARUBACEÆ.—A family of very bitter polypetalous trees or shrubs—the quassia family—having alternate pinnate leaves and small diœcious flowers in axillary panicles or racemes. It embraces 33 genera and 110 species, all natives of warm countries.

Official drug	Part used	Botanical name
Quassia	Wood	{ Picrasma excelsa { Quassia amara

FAMILY 43. ZYGOPHYLLACEÆ.—A family of polypetalous shrubs or herbs—the bean-caper family—having jointed branches, two-foliolate or pinnate stipulate leaves, and axillary peduncles bearing white, red,

or yellow flowers. It embraces 18 genera and 110 species, mainly tropical in distribution.

Official drug	Part used	Botanical name
Guiacum	Resin	{ Guiacum officinale { Guiacum sanctum

FAMILY 44. TERNSTROEMIACEÆ.—A family of polypetalous trees or shrubs—the tea or camellia family—having alternate simple leaves, and often large, showy, mostly five-parted flowers with numerous stamens. It embraces 41 genera and 310 species, nearly all natives of the tropics.

Official drug	Part used	Botanical name
Caffeina	Feebly basic principle	Thea chinensis

FAMILY 45. SAPINDACEÆ.—A family of polypetalous trees or shrubs —the soapberry family—having alternate, often evergreen, compound leaves, and small unsymmetrical odorless flowers with eight stamens. It embraces 122 genera, and 950 species, mainly tropical.

Official drug	Part used	Botanical name
Guarana	Paste of crushed seeds	Paullinia Cupana

FAMILY 46. STERCULIACEÆ.—A family of polypetalous shrubs, or trees—the cola-nut or sterculia family—having usually opposite, single, or three- to nine-foliate leaves and a variously shaped inflorescence of regular perfect flowers with frequently monadelphous stamens having two-celled anthers.

Official drug	Part used	Botanical name
Oleum theobromatis	Fixed oil	Theobroma Cacao
Unofficial		
Cola	Seed	Cola acuminata

FAMILY 47. UMBELLIFERÆ OR PARSLEY FAMILY.—A family of polypetalous herbs or shrubs characterized as follows:

Inflorescence, an umbel (simple or compound) of small flowers, each, with five petals and five stamens and ovary two-celled, inferior, calyx adnate to ovary.

Fruit, a cremocarp, consisting of two seed-like dry carpels or mericarps which often separate when fruit is ripe. Entire plants possess aromatic volatile oils.

Official drug	Part used	Botanical name
Anisum	Ripe fruit	Pimpinella Anisum
Fœniculum	Nearly ripe fruit	Fœniculum vulgare
Sumbul	Rhizome and roots	Undetermined
Carum	Fruit	Carum Carvi
Conium	Unripe fruit	Conium maculatum
Asafœtida	Gum resin	Ferula fœtida
Coriandrum	Ripe fruit	Coriandrum sativum

Unofficial

Angelica	Root	Angelica archangelica
Apium	Root	Apium petroselinum
Celery (fruit)	Fruit	Apium graveolens
Ammoniacum	Gum resin	Dorema Ammoniacum

FAMILY 48. ERYTHROXYLACEÆ.—Chloripetalous shrubs or trees with small zygomorphic flowers exhibiting a five-lobed calyx, five petals, 10 hypogynous stamens and a superior ovary; fruit a drupe. Indigenous to torrid and temperate zones.

Official drug	Part used	Botanical name
Coca	Leaves	{ Erythroxylon Coca { Erythroxylon Truxillense

FAMILY 49. VITACEÆ OR GRAPE FAMILY.—Chloripetalous shrubs with abundant watery sap, whose stems climb by means of tendrils opposite the leaves; flowers hypogynous; fruit a berry.

Official drug	Part used	Botanical name
Vinum Album	Fermented juice of fruit	Vitis vinifera
Vinum Rubrum	Fermented juice of fruit in presence of their skins.	Vitis vinifera

FAMILY 50. PUNICACEÆ, OR POMEGRANATE FAMILY.—Chloripetalous trees of small size with opposite ovate-lanceolate, entire leaves, scarlet receptacle, calyx and corolla; fruit an edible berry with hard rind.

Official drug	Part used	Botanical name
Granatum	Bark of stem and root	Punica Granatum

SUB-CLASS B. SYMPETALÆ GAMOPETALÆ

A division of dicotyledonous plants in which the flowers possess both calyx and corolla, the latter with petals more or less united into one piece.

FAMILY 1. RUBIACEÆ.—A large family of gamopetalous trees, shrubs, or herbs—the Madder family—with simple opposite or whorled leaves, connected by interposed stipules, and perfect, often dimorphous, flowers. It embraces 25 tribes, 375 genera, and 4500 species in all parts of the world.

Usually contain valuable alkaloids.

Official drug	Part used	Botanical name
Cinchona	Bark	Cinchona officinalis Cinchona Calisaya Cinchona L edgeriana, and hybrids
Cinchona Rubra	Bark	Cinchona succirubra
Ipecacuanha	Root	Cephælis Ipecacuanha Cephælis acuminata
Gambir	Extract	Ourouparia Gambir
Caffeina	Feebly basic substance	Coffea arabica

FAMILY 2. CONVOLVULACEÆ.—A large widely dispersed family of gamopetalous, chiefly climbing herbs, rarely shrubs or trees—the convolvulus or bindweed family—with alternate leaves, and showy pentamerous axillary flowers. It embraces about 36 genera and 870 species.

Contains milky juices.

Official drug	Part used	Botanical name
Jalapa	Tuberous root	Exogonium Purga
Scammonium	Gum resin	Convolvulus Scammonia

FAMILY 3. VALERIANACEÆ.—A family of gamopetalous herbs— the valerian family—having opposite exstipulate leaves and cymes of small often irregular flowers with stamens fewer than the corolla lobe, and inserted on its tube. There are nine genera and 275 species. *Valeriana*, the type genus, distinguished by its triandrous flowers, includes the common or official valerian.

Official drug	Part used	Botanical name
Valeriana	Rhizome and roots	Valeriana officinalis

FAMILY 4. SAPOTACEÆ.—A family of gamopetalous plants—the star-apple or the sapodilla family—being mainly trees or shrubs with milky juice, alternate leathery leaves, and large flowers with perfect stamens. It embraces 38 genera, and 400 species, all natives of the warmer countries.

An important resin-producing family.

Unofficial drug	Part used	Botanical name
Gutta-percha	Concrete exudation	Palaquium gutta

Family 5. GENTIANACEÆ.—A family of smooth annual or perennial gamopetalous herbs—the gentian family—with colorless bitter juice, opposite simple leaves, and showy, perfect, regular flowers. It is widely distributed over the world, especially in temperate regions, and embraces 49 genera, and about 575 species.

Official drug	Part used	Botanical name
Gentiana	Rhizome and roots	Gentiana lutea
Chirata	Entire plant	Swertia Chirayita

FAMILY 6. LOGANIACEÆ, THE LOGANIA FAMILY.—Tropical herbs, shrubs, or trees containing bitter principles, often poisonous. Allied to the milkweed and gentian families. Leaves entire, stipulate, opposite, inflorescence cymose, flowers, regular and four- to five-parted, fruit a two-celled berry or capsule.

Official drug	Part used	Botanical name
Gelsemium	Rhizome and roots	Gelsemium sempervirens
Spigelia	Rhizome and roots	Spigelia marilandica
Nux vomica	Seed	Strychnos nux vomica
Unofficial		
Cuara	Extract	Strychnos toxifera
Ignatia	Seed	Strychnos Ignatii

FAMILY 7. APOCYNACEÆ.—A family of gamopetalous herbaceous or woody plants—the dogbane family—mainly tropical or sub-tropical, with milky, mostly acrid juice, simple, entire, exstipulate leaves, and regular, five-parted flowers. It embraces 103 genera and 900 species.

Fruit, a pod containing many seeds which are often downy.

Official drug	Part used	Botanical name
Apocynum	Rhizome	Apocynum cannabinum and other species
Strophanthus	Seed	Strophanthus hispidus

FAMILY 8. ASCLEPIADACEÆ.—A large family of gamopetalous perennial herbs or shrubs—the milkweed family—erect or twining, having milky juice, leaves mostly opposite, five-parted umbellate flowers, stamens with the pollen cohering in waxy masses, and a fruit of two follicles. It embraces about 1300 species.

Unofficial drug	Part used	Botanical name
Asclepias	Root	Asclepias tuberosa
Condurango	Bark	Gonolobus condurango

FAMILY 9. CAPRIFOLIACEÆ.—A family of gamopetalous herbs shrubs, or rarely small trees—the honeysuckle family—mostly of the

northern hemisphere, having opposite lobed or odd-pinnate leaves, the inflorescence usually cymose with perfect regular or irregular flowers, and a baccate or drupaceous fruit. It includes 13 genera and about 200 species, the honeysuckle, viburnum, elder, etc.

Official drug	Part used	Botanical name
Viburnum Opulus	Bark	Viburnum opulus
Viburnum Prunifolium	Bark	{ Viburnum prunifolium { Viburnum lentago
Unofficial		
Sambucus	Flowers	Sambucus canadensis

FAMILY 10. SOLANACEÆ.—A family of gamopetalous, frequently narcotic, poisonous plants—the nightshade family—having colorless juice, alternate simple leaves, regular pentamerous and pentandrous flowers and many seeds. It embraces 72 genera, and 1750 species, found in all warm countries, particularly America. *Solanum*, the type genus, includes *S. tuberosum*, the cultivated potato; *S. Melongena*, the egg-plant; *S. nigrum*, the black nightshade; *S. Dulcamara*, the bittersweet; *S. Carolinense*, the Horse Nettle.

Official drug	Part used	Botanical name
Belladonnæ Folia	Leaves	Atropa Belladonna
Belladonnæ Radix	Root	Atropa Belladonna
Stramonium	Leaves	Datura Stramonium
Hyoscyamus	Leaves and flower tops	Hyoscyamus niger
Scopola	Rhizome	Scopola Carniolica
Capsicum	Fruit	Capsicum fastigiatum
Unofficial		
Dulcamara	Twigs	Solanum dulcamara
Duboisia	Leaves	Duboisia myoporoides
Tabacum	Leaves	Nicotiana tabacum

FAMILY 11. CAMPANULACEÆ.—A family of gamopetalous herbs—the bellwort family—of northern temperate regions, with alternate simple leaves and regular blue or white bell-shaped five-parted flowers, embracing 53 genera (including the *Lobeliaceæ*) and a thousand species.

Official drug	Part used	Botanical name
Lobelia	Leaves and flowering tops	Lobelia inflata

FAMILY 12. ERICACEÆ.—A family of gamopetalous trees, shrubs, or perennial herbs—the heath family—with commonly alternate, undivided, often evergreen leaves, variously shaped clusters of symmet-

rical tetramerous or pentamerous flowers, and capsular, baccate or drupaceous fruit.   They are natives of temperate or cold climates.

Leaves have a bitter astringent taste due to glucosides.   Blossoms bell-shaped or wen-shaped.

Official drug	Part used	Botanical name
Chimaphila	Leaves	Chimaphila umbellata
Uva Ursi	Leaves	Arctostaphylos Uva Ursi
Oleum Gaultheriæ	Volatile oil	Gaultheria procumbens

FAMILY 13.   OLEACEÆ.—A family of gamopetalous erect or climbing shrubs, trees, or rarely herbs—the olive family—with opposite, simple or pinnate leaves and perfect or unisexual flowers with four-lobed calyx, four-cleft corolla, and two or rarely four free stamens.   It embraces 19 genera, and about 300 species, distributed over the warm or temperate regions of the world.

Official drug	Part used	Botanical name
Oleum Olivæ	Fixed oil	Olea europea
Manna	Saccharine exudate	Fraxinus ornus

FAMILY 14.   SCROPHULARIACEÆ.—A family of gamopetalous plants —the figwort family—chiefly herbs with various forms of leaves and inflorescence, the flowers distinguished by having a persistent five-lobed calyx and a two-lipped corolla with four didynamous stamens, and often one staminode inserted on its tube, and the fruit a two-celled, usually many-seeded capsule with axile placentæ.   It embraces 166 genera, and more than 2000 species.   Contains bitter, acrid, poisonous principles.

Official drug	Part used	Botanical name
Digitalis[1]	Leaves	Digitalis purpurea
Leptandra	Rhizome and roots	Veronica virginica
Unofficial		
Verbascum	Leaves	Verbascum thapsus

FAMILY 15.   LABIATÆ OR MINT FAMILY.—A cosmopolitan family of sympetalous herbs, rarely shrubs, with quadrangular stems, opposite or whorled aromatic leaves, and usually thyrsoid or verticillate clusters of flowers, each with a two-lipped corolla, didynamous or diandrous stamens, and a four-lobed ovary.   All of the members of this family are rich in volatile oils.

Official drug	Part used	Botanical name
Salvia	Leaves	Salvia officinalis
Scutellaria	Dried plant	Scutellaria lateriflora
Marrubium	Leaves and flowering tops	Marrubium vulgare
Hedeoma	Leaves and flowering tops	Hedeoma pulegioides
Mentha Viridis	Leaves and flowering tops	Mentha spicata
Mentha Piperita	Leaves and flowering tops	Mentha piperita
Oleum Thymi	Volatile oil from leaves and flowering tops	Thymus vulgaris
Oleum Rosmarini	Volatile oil from fresh flowering tops	Rosmarinus officinalis
Oleum Lavendulæ Florum	Volatile oil from fresh flowering tops	Lavendula officinalis
Unofficial		
Melissa	Leaves and tops	Melissa officinalis
Origanum	Herb	Origanum majorana

FAMILY 16. STYRACEÆ.—A family of gamopetalous trees or shrubs —the storax family—having alternate simple leaves and usually white racemed flowers with a corolla of four to eight more or less united petals. It embraces seven genera and 235 species, natives of all parts of the world.

Official drug	Part used	Botanical name
Benzoinum	Balsamic resin	Styrax Benzoin

FAMILY 17.   COMPOSITÆ.—The largest family of plants embracing 835 genera, and over 10,000 species. A family of gamopetalous herbs, shrubs and rarely trees found in all parts of the world, having their flowers in a head or capitulum on a common receptacle, surmounted by an involucre, with five (rarely four) stamens inserted on the carolla, their anthers, syngenesious. Calyx tube crowned by a pappus in the form of bristles, teeth or scales, etc. Corolla either ligulate or tubular. In the perfect flowers a two-cleft style is present. Fruit, an akene. The plants of this family contain inulin, a substance isomeric with starch.

Official drug	Part used	Botanical name
Anthemis	Flower head	Anthemis nobilis
Arnica	Flower head	Arnica montana
Matricaria	Flower head	Matricaria Chamomilla
Calendula	Ligulate florets	Calendula officinalis
Lappa	Root	Arctium lappa
Pyrethrum	Root	Anacyclus Pyrethrum
Taraxacum	Root	Taraxacum officinale

Eupatorium	Leaves and flowering tops	Eupatorium perfoliatum
Grindelia	Leaves and flowering tops	Grindelia robusta Grindelia squarrosa
Lactucarium	Concrete milk juice	Lactuca virosa
Santonica	Unexpanded flower heads	Artemisia pauciflora
Oleum Erigerontis	Volatile oil	Erigeron canadense
Unofficial Pyrethri Flores	Flower heads	Chrysanthemum roseum Chrysanthemum carneum
Carthamus	Florets	Carthamus tinctorius
Cichorium	Root	Cichorium intybus
Inula	Root	Inula Helenium
Absinthium	Leaves and flowering tops	Artemisia Absinthium
Achillea	Leaves and flowering tops	Achillea millefolium
Tanacetum	Leaves and flowering tops	Tanacetum vulgare

FAMILY 18.   HYDROPHYLLACEÆ.—Herbaceous, shrubby, or arborescent plants containing a watery, insipid juice and further characterized by having hairy and toothed pinnately compound leaves; scorpoid inflorescences, and two-valved fruits.

Official drug	Part used	Botanical name
Eriodictyon	Leaves	Eriodictyon californicum

FAMILY 19.   CUCURBITACEÆ.—A natural family of usually succulent tendril-bearing dicotyledonous herbs—the gourd family—with climbing or prostate stems, simple plamately veined alternate leaves, monœcious or diœcious, rarely gamopetalous flowers, and a large, fleshy, usually three-celled fruit.   It embraces 86 genera and about 630 species, mostly found in the tropics.

Official drug	Part used	Botanical name
Pepo	Seed	Cucurbita Pepo
Colocynthis	Peeled dried fruit	Citrullus Colocynthis
Elaterinum	Neutral principle	Ecballium Elaterium

SUBDIVISION II.—THE GYMNOSPERMS

The Gymnosperms comprise an ancient and historic group of plants which were more numerous in the Paleozoic and Carboniferous periods than now.   They differ from the Angiosperms mainly in their seeds being exposed.   Most of their number are evergreens, retaining their leaves throughout the year, important exceptions being the Larches which drop their foliage upon the advent of winter.

The groups still extant are the Cycads or Fern Palms, the Gnetums, the Ginkgos, and the Conifers.   Of these the Conifers are of most pharmaceutic importance. This, the largest group of Gymnosperms, includes the pines, firs, spruces, hemlocks, junipers, balsams, cedars, and arbor vitæ. The following Gymnospermous plants yield products of pharmaceutic and medicinal value:

Fig. 38.—Inflorescences of the pine. 1. Terminal twig; 2. ovulate cone; 3. staminate cone; 4. two-year-old cone. (*From Hamaker.*)

Botanical name	Products
Pinus strobus	White pine bark
Pinus palustris	
Pinus glabra	
Pinus echinata	Turpentine, rosin and tar
Pinus tæda and	
other species of Pinus	
Abies balsamea	Balsam of fir
Larix decidua	Venice turpentine
Picea excelsa (Abies excelsa)	Burgundy pitch
Tsuga occidentalis	Volatile oil
Juniperus Oxycedrus	Oil of cade
Juniperus communis	Juniper berries and volatile oil
Juniperus Sabina	Tops and volatile oil of savin
Callitris quadrivalvis	Sandarac
Pinus sylvestris	Volatile oil

Family 1. Pinaceæ.—Old name Coniferæ. The pine family. (Cone-bearing family.) Trees or shrubs, with resinous juice, mostly awl-shaped or needle-shaped leaves, and monœcious or rarely diœcious flowers in catkins, destitute of calyx or carolla. Three sub-families, Abietineæ, or proper pine family; Cupressineæ, or cypress family; and Taxineæ, or yew family. All are evergreen excepting the Larches.

Official drug	Part used	Botanical name
Terebinthina	Concrete oleoresin	Pinus palustris and
Resina	Resin	other species
Pix Liquida	Destructive distillate product	Other species
		Other species
Terebinthina Canadensis	Liquid oleoresin	Abies balsamea
Sabina	Tops	Juniperus Sabina
Oleum Cadinum	Oily product	Juniperus Oxycedrus
Oleum Juniperi	Volatile oil	Juniperus communis
Unofficial		
Pix Burgundica	Resinous exudate	Abies excelsa
Sandaraca	Resinous exudate	
	Resinous exudate	Callitris quadrivalvis
Dammar	Resinous exudate	Agathis loranthifolia
Pix Canadensis		Tsuga canadensis
Succinum (amber)	Fossil resin	Pinites succinifer
Terebinthina Veneta	Oleoresin	Larix europea
Juniperus	Fruit	Juniperus communis

BIBLIOGRAPHY

Le Maout and Decaisne's "Descriptive and Analytical Botany."

"Lehrbuch der Botanik," by Strasburger, Noll, Schenck and Schimper.

Bastin's "College Botany."

Macfarlane's "Lectures on the Comparative Morphology and Taxonomy of the Angiospermia".

Kramer's "Botany and Pharmacognosy."

Steven's "Plant Anatomy."

Hamaker's "Outlines of Biology."

"A Text Book of Botany," vol. I, by Coulter, Barnes, & Cowles.

"Elements of Botany and Pharmacognosy," by Mary L. Creighton.

Die Naturlichen Pflanzenfamilien von A. Engler und K. Prantl.

Warming's "Systematic Botany," translated by Potter.

INDEX

Frond, 67
Fruit, 50–53
   classification, 50–53
   definition, 50–53
   dehiscence, 50–53
   distribution, 50–53
   structure, 50–53
Fruticose, 21
Fucus, 59
Funaria, 65
Fundamental tissue, 5
Fungi, 59

Galla, 73
Gambir, 86
Gameotophytes, 64
Gametophyte, 49, 65
   male, 49
   female, 49
Gamopetalæ, 85
Garcinia Hanburii, 81
Garlic, 71
Gasteromycetes, 63
Gaultheria procumbens, 89
Geaster, 63
Geradia, 16
Geraniaceæ, 80
Geranium Family, 80
   maculatum, 80
Germination, 50
Gelsemium sempervirens, 87
Gemma, 64
Gentianaceæ, 4, 87
Gentiana lutea, 87
Gentianose, 4
Genus, 2
Geological Botany, 1
Ginger Family, 72
Ginkgos, 91
Gloeocapsa, 56
Glomerule, 39
Glucosides, 4
Glycyrrhiza glabra, 78
   glandulifera, 78
Gnetums, 91
Gonolobus condurango, 87
Goosefoot Family, 75
Gossypii Cortex, 81
Gossypium Purificatum, 81
   herbaceum, 81
Gourd Family, 91
Grain, 52
Graminaceæ, 70
Granatum, 85
Grass Family, 70
Grindelia robusta, 91]
   squarrosa, 91
Guard cells, 6

Guarana, 84
Guiacum officinale, 84
   sanctum, 84
Gutta-percha, 86
Guttiferæ, 81
Gymnosperms, 2, 23, 91
Gynœcium, 46
   Angiospermous, 46
   definition, 46
   Gymnospermous, 46
   Structure, 46

Hagenia abyssinica, 77
Hairs, plant, 14
   root, 15
Hamamelidaceæ, 80
Hamamelis virginiana, 80
Hamamelidis Cortex, 80
   Folia, 80
Hard bast, 5, 8
Hybrid, 2
Haustoria, 16, 59
Hæmatoxylon campechianum, 78
Heath Family, 88
Hedeoma pulegioides, 90
Helleborus niger, 78
Hemlocks, 92
Hepaticæ, 64
Herb, 21
Hesperidium, 53
Hevea, 76
Histology of dicotyl stem, 23
   root, 18
   of monocotyl stem, 29
   root, 17
Honey Locust, 21
Honeysuckle Family, 87
Hordeum distichum, 70
Horse Nettle, 88
   Horsetails, 66
Hydrastis, 78
Hydrophyllaceæ, 91
Hydrophilous, 49
Hydropteridineæ, 69
Hyoscyamus niger, 88
Hydnaceæ, 63
Hydnum, 63
Hymenium, 63
Hymenomycetes, 63
Hypha, 59
Hypnum, 65
Humulus lupulus, 76

Ignatia, 87
Illicium verum, 76
Indeterminate Inflorescence, 37–39
   definition, 37–39
   kinds, 37–39

Made in the USA
Monee, IL
08 July 2026